AF684232

Build Your Own Test Equipment

By Carl Bergquist

Build Your Own Test Equipment

By Carl Bergquist

PROMPT® PUBLICATIONS

©1998 by Howard W. Sams & Company

PROMPT© Publications is an imprint of Howard W. Sams & Company, A Bell Atlantic Company, 2647 Waterfront Parkway, E. Dr., Indianapolis, IN 46214-2041.

All rights reserved. No part of this book shall be reproduced, stored in a retrieval system, or transmitted by any means, electronic, mechanical, photocopying, recording, or otherwise, without written permission from the publisher. No patent liability is assumed with respect to the use of the information contained herein. While every precaution has been taken in the preparation of this book, the author, the publisher or seller assumes no responsibility for errors or omissions. Neither is any liability assumed for damages resulting from the use of information contained herein.

International Standard Book Number: 0-7906-1130-9
Library of Congress Catalog Card Number: 97-68184

Acquisitions Editor: Candace M. Hall
Editor: Natalie F. Harris
Assistant Editors: Pat Brady, Loretta Leisure
Editorial Support: Shey Query
Typesetting: Natalie Harris
Indexing: Natalie Harris
Cover Design: Kelli Ternet
Illustrations: Carl Bergquist
Graphics Conversion: Terry Varvel
Additional Materials: Schematics, parts list and operational description of the CA-1 capacitance adapter kit courtesy of Electronic Rainbow, Inc.

Trademark Acknowledgments:
All product illustrations, product names and logos are trademarks of their respective manufacturers. All terms in this book that are known or suspected to be trademarks or services have been appropriately capitalized. PROMPT© Publications, Howard W. Sams & Company, and Bell Atlantic cannot attest to the accuracy of this information. Use of an illustration, term or logo in this book should not be regarded as affecting the validity of any trademark or service mark.

PRINTED IN THE UNITED STATES OF AMERICA

9 8 7 6 5 4 3 2 1

Contents

Introduction

During my years of electronics exploration, I have always had a fascination with test equipment. Many of the articles I've written in the past are dedicated to just that subject, relentlessly bearing economy in mind.

Not only does test equipment help you out, building it can carry you a long way toward understanding electronics. In addition, the procedures allow you to customize the equipment to your actual needs, as opposed to your "needs" as perceived by an OEM (original equipment manufacturer).

I'm not knocking the available commercial gear. The vast majority of it is very good. However, often the engineers assigned to developing the gear tend to overdo the process; that is, too many features are either rarely used or of little use at all.

On the other side of the OEM coin is the equipment that doesn't have the features you want. You are left to your own devices in finding ways to supplement what does exist. Not the best way to go by any means.

So, I have compiled this text in an effort to help you out. Of course, I've helped myself out along the way, but this is just one of the perks of writing this type of book. I have also made every attempt possible to keep construction costs at a minimum. I mean, who enjoys confronting some stingy banker or "Big Louie the Knife" in order to finance a project? Not I. So, second or third mortgages on the house or car are not usually necessary when building these devices. I won't say "never", as each individual's financial condition is different; but I do hope such drastic measures will not be needed.

Anyway, within the pages of this book you will find twelve of the most pragmatic testing devices ever conceived by mankind. (*Photos A-1* and *A-2*) They are useful, too. I wouldn't lie to you about something like that. Honest I wouldn't!

The first chapter describes a prototyping lab that will greatly improve both design speed and efficiency. The remaining chapters detail such items as a bench power supply, IC tester, digital function generator, bench multimeter, frequency counter and more.

Photo A-1. A view of all twelve complete devices in their unpowered or OFF state.

Photo A-2. All twelve completed projects in their ON state.

Each and every device is designed to be highly practical and space conscious. I know all too well how tight a workbench can get. Many of these gems stack nicely, thus conserving precious bench real estate.

As always, I have made every effort to use commonly-available components. Anything special includes a known supplier in the parts list, so you won't be up all night on the Internet hunting the stuff down (and calling me ugly names in the process). Seriously, there are few circumstances more frustrating (not to mention infuriating) than trying to locate a bizarre chip for a project you want to build. Such a scenario can understandably invoke a refresher course in four-letter words.

I have also tried to include as many construction tips as possible, for those pesky little anomalies that crop up in every project. If I found them, you will hear about them. Part of my obligation is to make these projects bugproof whenever feasible.

On a final note, I always encourage modification (whenever desired) to my designs. This is unquestionably one of the best ways to learn. I'm what you might call a nu.... uh, *advocate* of printed circuit boards (PCBs); thus all of the projects include PCB foil patterns. However, this doesn't necessarily mean you have to use them.

Many of these devices will work just fine if perfboard and point-to-point wiring is employed. Don't be afraid to use techniques that you find comfortable. The PCB patterns, cases and many other suggestions are just that—suggestions. As the saying goes, "Nothing is etched in stone."

I'm sure you have developed techniques that, in the past, have proven reliable and functional. I have only included my comments as possible guides or assistance in making this experience a fond one.

So, here it is. Twelve test bench projects in one book. If you add all of these devices to your bench, you will have one heck of an electronics prototyping and experimental lab. This was one of my primary parameters while planning this text.

If you are like me, you have transistors, capacitors, ICs and all the rest flowing in your veins. The field takes dedication, but it can also be captivating. With the right equipment, the learning process becomes faster and far more pleasant. This hobby is one you can really learn to love.

Do some learning; but most of all, have fun!

Chapter 1
The Ultimate Prototype Lab

Introduction

This is one project you will NOT be sorry you built. Once completed, the prototype lab will allow for easy breadboarding of virtually all of the circuits you want to test or develop.

In an effort to avoid the jumbled maze of hookup wires usually associated with breadboard projects, the lab panel mounts all of the cumbersome components (such as potentiometers and switches) neatly around the solderless breadboards.

Each component has a set of single in-line pin (SIP) sockets of the appropriate number, providing access to the component. Additionally, an audio signal generator, audio amplifier, logic probe, digital voltmeter and power supply are included in the design. All are highly useful when prototyping circuits, especially original configurations. Again, SIP sockets or banana jacks handle the input/output duties.

One nice feature of this project is that YOU decide how complex or how simple it will be. In this chapter, the system I built will be discussed, and hopefully each element will be explained clearly. However, you may well want to tailor your unit for a different approach or purpose.

Another pleasant surprise of this project is that many of the items required are likely lingering in your junk box. Hence, construction costs will not require you to get a bank loan or second mortgage, or hold up a stranger on the street.

PROTOTYPE LAB PARTS LISTS

AUDIO SIGNAL GENERATOR
SEMICONDUCTORS

U1	LM555 Timer/Oscillator IC

RESISTORS

R1	1,000 Ohm 1/4 Watt Resistor
R2	1,000,000 Ohm Panel Potentiometer

CAPACITORS

C1	10 Microfarad Electrolytic Capacitor
C2	0.047 Mylar Capacitor

OTHER COMPONENTS

S1	SPDT Slide Switch

AUDIO AMPLIFIER
SEMICONDUCTOR

U1	LM386N-3 Low Voltage Audio Amplifier (500mw/9v)

RESISTORS

R1	5,000 Ohm Panel Potentiometer
R2	10 Ohm 1/4 Watt Resistor

CAPACITORS

C1	0.05 Microfarad Mylar Capacitor
C2	220 Microfarad Electrolytic Capacitor

LOGIC PROBE
SEMICONDUCTORS

U1	74LS04 Hex Inverter
LED1	Red Jumbo T-1 3/4 Light Emitting Diode
LED2	Green Jumbo T-1 3/4 Light Emitting Diode

RESISTOR

R1 220 Ohm 1/4 Watt Resistor

DIGITAL VOLTMETER
SEMICONDUCTORS

U1 ICL7107 3 1/2 Digit LED A/D Converter
DSP1, 2 MAN6710 Dual Digit CA LED Displays (or equivalent)

RESISTORS

R1 47,000 Ohm 1/4 Watt Resistor
R2 1,000,000 1/4 Watt Resistor
R3 1,000 Ohm PCB Potentiometer
R4 22,000 Ohm 1/4 Watt Resistor
R5, 8, 9 100,000 Ohm 1/4 Watt Resistors
R6 470 Ohm 1/4 Watt Resistor
R7, 10 10,000,000 Ohm 1/4 Watt Resistors

CAPACITORS

C1 0.22 Microfarad Mylar Capacitor
C2 0.47 Microfarad Mylar Capacitor
C3 0.01 Microfarad Mylar Capacitor
C4 0.1 Microfarad Mylar Capacitor
C5 100 Picofarad Ceramic Disk Capacitor

OTHER COMPONENTS

SW1, 2 DPDT Slide Switches
J1, 2 Banana Jacks

POWER SUPPLY
SEMICONDUCTORS

U1 7805T Positive 5 Volt Regulator
U2 7905T Negative 5 Volt Regulator
U3 7812T Positive 12 Volt Regulator
U4 7912T Negative 12 Volt Regulator

U5	LM317T Positive Adjustable Voltage Regulator
LED1	Green Jumbo T-1 3/4 Light Emitting Diode
BR1-3	200 PIV, 1.5 Amp Bridge Rectifiers (KBP02, or Equiv)

RESISTORS

R1, 3	220 Ohm 1/4 Watt Resistors
R2	5,000 Ohm Panel Potentiometer

CAPACITORS

C1, 2	470 Microfarad, 16 Volt Electrolytic Capacitors
C3, 4	1 Microfarad, 50 Volt Tantalum Capacitors
C5, 6	470 Microfarad, 25 Volt Electrolytic Capacitors
C7, 8	4.7 Microfarad, 35 Volt Tantalum Capacitors
C9	1,000 Microfarad, 35 Volt Electrolytic Capacitor
C10	10 Microfarad, 35 Volt Tantalum Capacitor

OTHER COMPONENTS

T1	120 VAC Primary, 3 Secondary Power Transformer. Secondaries in the 10 VAC, 18 VAC, 29 VAC Range.
S1	SPST Power Switch (125 Volt)
P1	AC Outlet Plug/Cord Assembly
HS1-5	Heat Sinks for All Regulators

PROTOTYPE LAB HARDWARE

1	0-1 Milliamp Analog Panel Meter Movement
1	MAN6740 Common Cathode LED Display
1	4 Digit Common Anode Multiplexed LED Array
1	8 mm Light Emitting Diode (any color)
2	Banana Jacks (AC Output)
77	Single In-Line Pin (SIP) Sockets
1	1,000 Ohm Panel Potentiometer
1	10,000 Ohm Panel Potentiometer
1	100,000 Ohm Panel Potentiometer
1	1,000,000 Ohm Panel Potentiometer
1	1-1/2" 8 Ohm Speaker
1	DPST Slide Switch (Signal Gen/Amp power)
1	2 Position Barrier Strip (Adj. Volt Out)
1	6 Position Barrier Strip (+/- 5, 12 V. Out)

1	SPST Toggle Switch (DVM "ON")
1	SP6T Rotary Switch
1	Push Button Momentary Contact Switch
1	DPDT Rocker Switch
8	3/4" Instrument Knobs
3	830 Tie Point Solderless Breadboards
1	11" x 14" x 1-1/4" Plastic Box Frame (Case)

Paint, Solder, Hardware, Dry Transfer Lettering, Rubber Feet, PCB Materials, Clear Red Plastic (DVM Facing), Etc.

In essence, this project should be fun. It permits wide latitude for modification, and the finished product will see endless service. So, without further delay, let's get our hands dirty and build this rascal.

Theory

As stated earlier, the primary purpose of the lab is to make prototyping less complicated. Solderless breadboards are great, but some components don't take kindly to them. Consequently, if your projects are like mine used to be, there are lead wires running in every direction imaginable. Not only is this sloppy, but it lends itself to trouble. Invariably some of the leads want to come loose, particularly if you have to move the breadboard any distance. Also, hookups have a nasty habit of becoming tangled and twisted; the whole thing can end up a mess.

One day I had enough of all this and decided there had to be a better way. This astounding revelation gave birth to the prototype lab. At least it got me thinking about it. So, let's talk about what the lab has to offer.

A major portion of the unit is occupied by solderless breadboards. I used three 2-1/4" (57 mm.) by 6-1/2" (166 mm.) boards, each with two distribution strips (200 tie points) and one terminal strip (630 tie points).

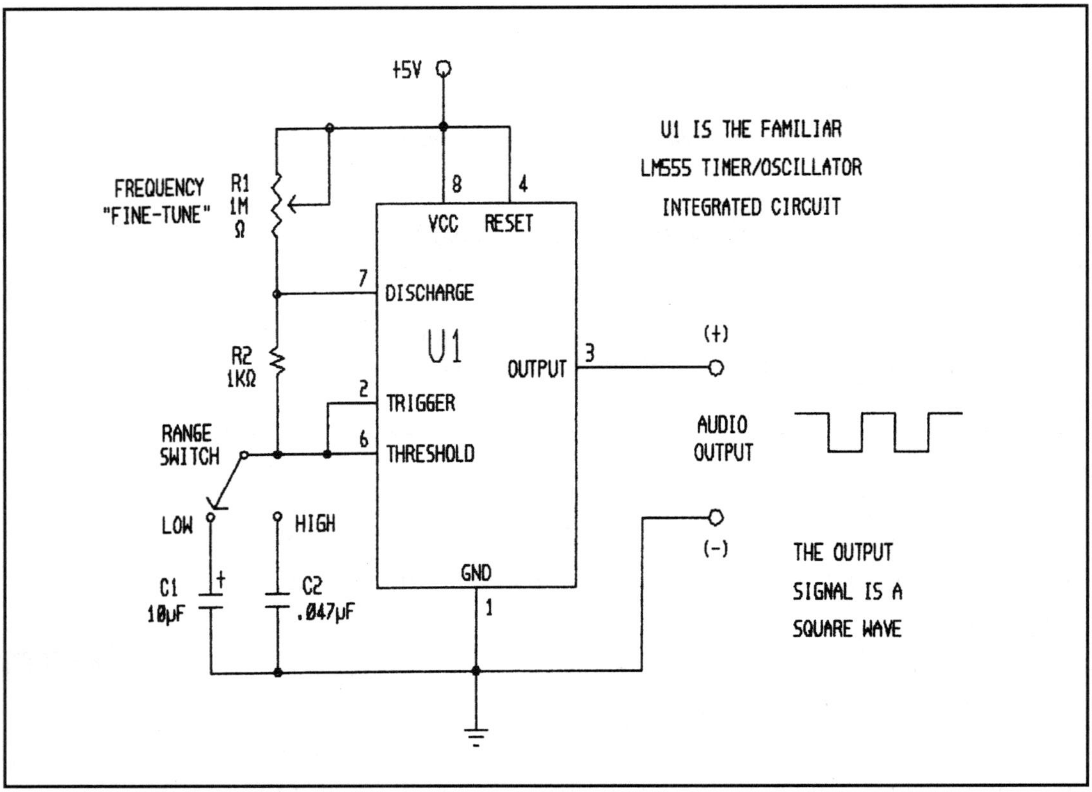

Figure 1-1. An audio signal generator.

The distribution strips carry power down the full length of the unit, while the terminal section provides an area for circuit components. With three such groupings, you have a total of 1,890 tie points for parts and 600 for the power connection. This will handle all but the most elaborate designs; however, if you feel it's not enough, you can add another board or two.

The breadboards are available from most suppliers, but I wouldn't recommend buying the cheapest units. You will have thousands of components coming in and going out of the tie points; thus, it is better to splurge a little and get good quality.

As for components, I added four potentiometers (1K, 10K, 100K and 1M), three switches (pushbutton momentary contact, SP6T rotary and

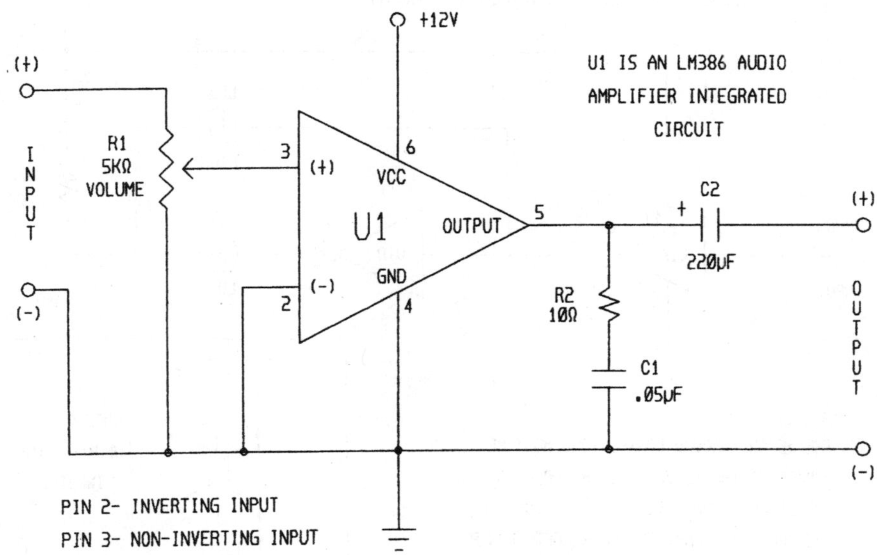

Figure 1-2. Audio amplifier.

DPDT rocker), a 0- to 1-milliamp analog meter, an LED, a speaker and two seven-segment light-emitting diode (LED) arrays.

The potentiometers, switches, LED, speaker and meter speak for themselves, while the LED displays are commonly-employed configurations. One is a four-digit, common-anode (CA) multiplexed unit, and the second handles either common anode or common cathode (CC) dual-digit displays.

The secret to this second arrangement is that the MAN 6710 CA and MAN 6740 CC dual displays have the same pinout. Thus, either unit will function when plugged into the socket. Incidentally, many of the dual-digit displays manufactured by other companies also follow the MAN pinout, so these displays don't necessarily have to be 6710s or 6740s. Try to use what you have on hand.

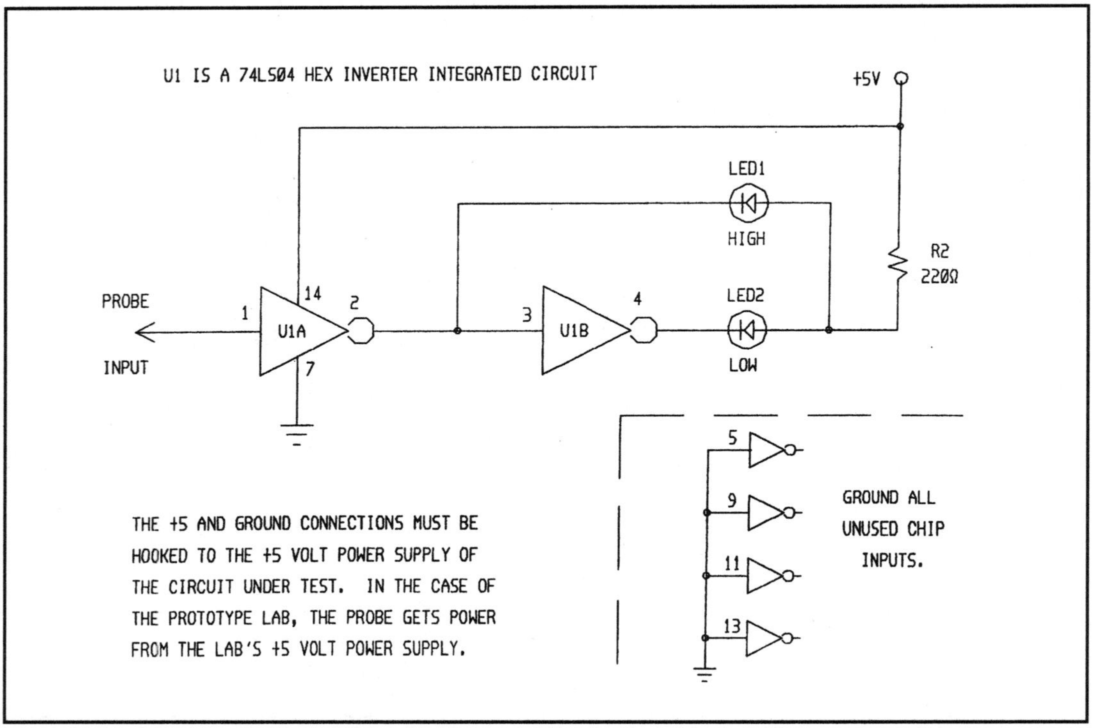

Figure 1-3. Logic probe.

Regarding the electronic circuits, let's first look at the signal generator. This is a simple LM555 timer astable configuration that produces two levels of signal, high and low. As seen in *Figure 1-1*, capacitors C1 and C2, in conjunction with potentiometer R1, determine the frequency of the oscillator. They are chosen to cover the spectrum from a one-second pulse all the way to the top end of the audio range.

Switch S1 selects either a 10-microfarad or 0.047-microfarad capacitor, with the latter producing the higher frequencies. R1 is used to fine tune the signal to what you need; the output will be of the square wave variety.

The audio amplifier is a standard application of the LM386, as seen in *Figure 1-2*. The potentiometer controls volume by limiting the strength of the input signal. There is nothing unusual about this circuit.

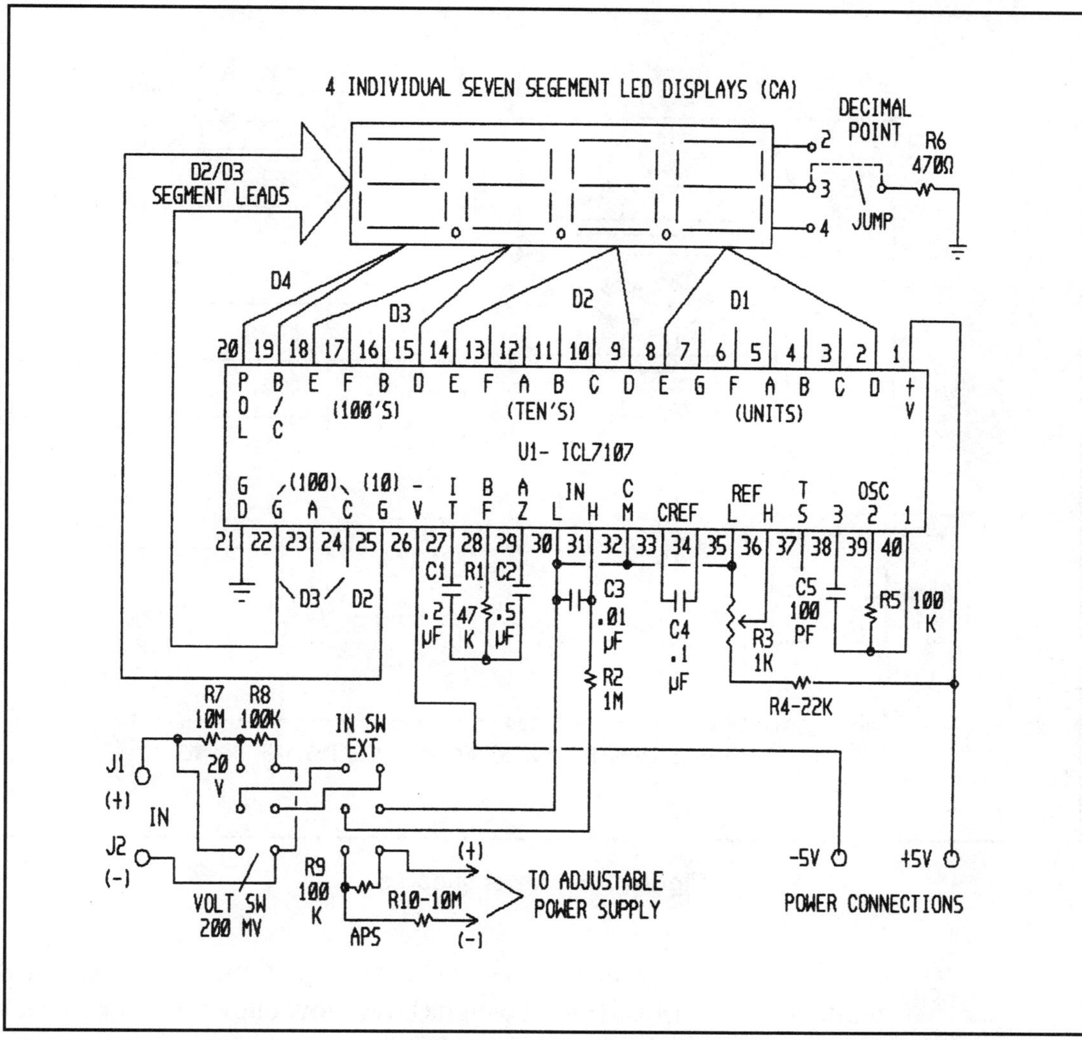

Figure 1-4. 7107 digital voltmeter.

As for the logic probe, it utilizes two gates of a 74LS04 hex inverter chip. *Figure 1-3* shows the schematic. As can be seen, the probe input is fed to the first inverter's (U1A) input. U1A's output is connected to both the cathode of LED1 and the input of the second inverter (U1B). In this fashion, a high input is inverted to a low which accommodates LED1's cathode, and turns it on. So, if LED1 is illuminated, the input is high, or floating.

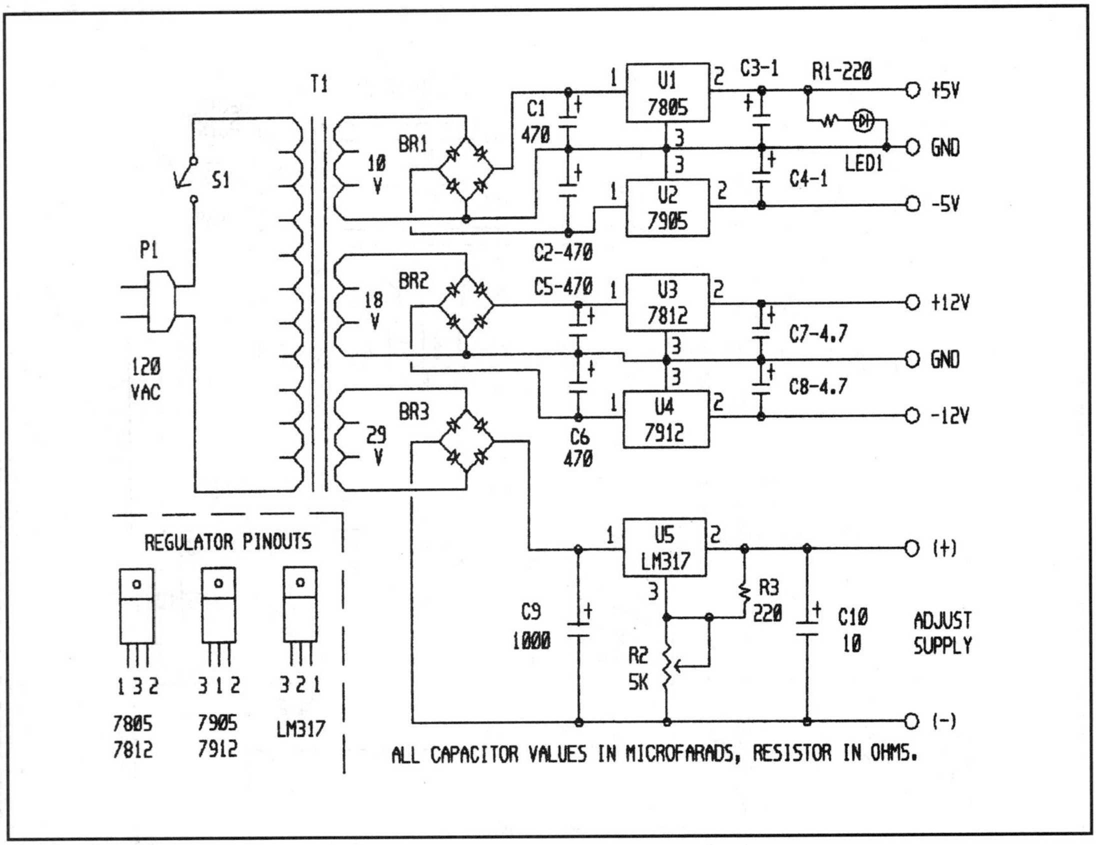

Figure 1-5. Power supplies.

U1B's output is connected to LED2's cathode; thus, when a signal is high, the output of U1B is also high (U1A's low output is inverted back to a high by U1B). This being of incorrect polarity for the cathode, LED2 stays dark. However, when the probe detects a low, U1A turns it into a high which is converted back to a low by U1B. This now satisfies LED2's needs, and it lights up to beat the band!

One final note on the logic probe. Since only two of the six gates are employed, it is prudent to ground the inputs of all unused gates (U1C-F). You could wire the unused inputs/outputs to SIP sockets for external access; but if you take that path, be sure to ground the input of any gate not in use. This prevents excess noise and power consumption.

The digital voltmeter is a 200-millivolt configuration of the Harris/Intersil ICL7107 IC (*Figure 1-4*). If you have read any of my past work, you are probably familiar with my admiration for this chip. It has been around for a while, but is still one of the best ways to read and digitally display direct current (DC) voltages.

The voltage is applied to input pins 30 and 31, processed by the 7107 and displayed on common anode LED arrays. All counting, decoding and display driving is handled by the chip. For this reason, the 7107 or its liquid crystal display (LCD) cousin, the 7106, is the basis for a good many commercial digital multimeters (DMMs).

For the prototype lab, the basic circuit is used, with the lab's internal supply furnishing the split power requirements (the 7107 needs +/- 5 volts). Once calibrated, which I will cover in the construction section, the meter will accurately read DC voltages from 0 to 200 millivolts.

In the interest of greater convenience, switch S2 introduces a voltage divider, allowing the meter to read from 0 to 20 volts. Switch S3, on the other hand, governs the input. The top position selects a pair of panel-mounted banana jacks, while the bottom position taps into the adjustable power supply's output. Again, through a divider circuit, this lets you monitor the supply's output voltage.

Regarding the power supplies, *Figure 1-5* reveals three separate standard linear-regulated systems. The first section provides the positive/negative 5 volts, number two furnishes plus/minus 12 volts, and the last segment is the positive adjustable supply. You'll learn to love this last one. In each case, the input alternating current (AC) is rectified, filtered and introduced to the input of the respective regulator IC. The output is again filtered by a small capacitor, which also prevents oscillation and reduces transient interference.

The transformer I used was robbed from a deceased audio system, and has three separate secondaries (10V, 18V and 29V). I can't supply a specific part number, but transformers of this sort are available from a number of resources. Many surplus dealers handle a wide variety of

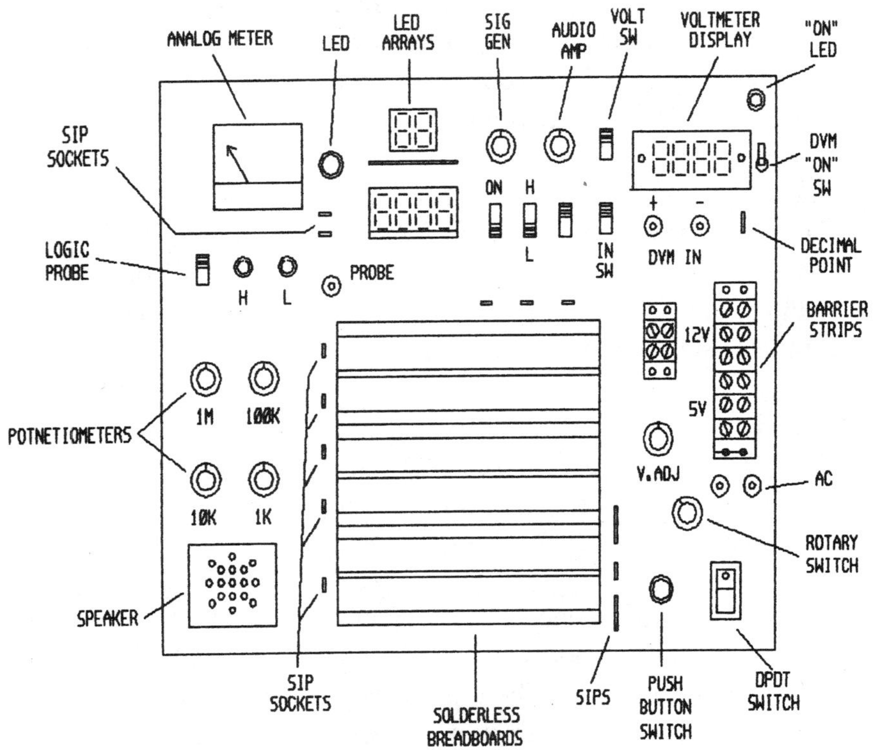

Figure 1-6. The prototype lab layout.

unusual units, and as mentioned, audio component systems and old VCRs are highly likely to contain multi-secondary power transformers. The secondaries only have to be a couple of volts above the regulator's rating (say 7 or 8 volts for the 5-volt supply). Current should be a minimum of 500 milliamps, with 1 or 2 amps preferred. The regulators will handle up to 1.5 amps, but this is more than you will probably need for most projects.

Barrier strips provide panel access to each of the five individual voltages, and a pair of banana jacks furnish AC power from the 29-volt transformer secondary. The AC comes in handy when rectification is part of a design, or a circuit otherwise requires an alternating current.

That pretty much covers the nuts and bolts of the prototyping lab or station, at least as far as the electronics are concerned. Next let's put the whole thing together.

Construction

Here again, there is plenty of room for modification, innovation or anything else you want. I chose an 11" (280 mm) by 14" (356 mm) by 1-1/4" (32 mm) plastic box frame as the case because plastic is easy to work with. However, virtually any enclosure you have will work just as well.

If you decide to go my route, these frames are readily available in discount and craft stores, and from some framing shops. They are relatively inexpensive (about $5 for an 11" x 14" frame), and the plastic is easy to drill and cut. One word of caution, however: this type of plastic is fairly brittle, so be careful not to apply too much pressure as it will crack.

Figure 1-6 illustrates the layout I employed for the lab, and can act as a guide for your unit. In this configuration, everything fits neatly with plenty of room for all panel components.

The first step is to mark and drill holes for each of the openings needed. For rectangular apertures, such as the meter movement and digital voltmeter display, drill small holes at each corner and use a hobby saw to

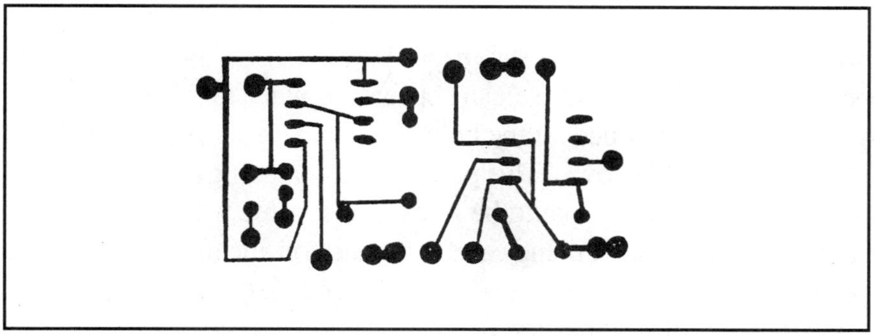

Figure 1-7. Audio generator/amplifier foil pattern.

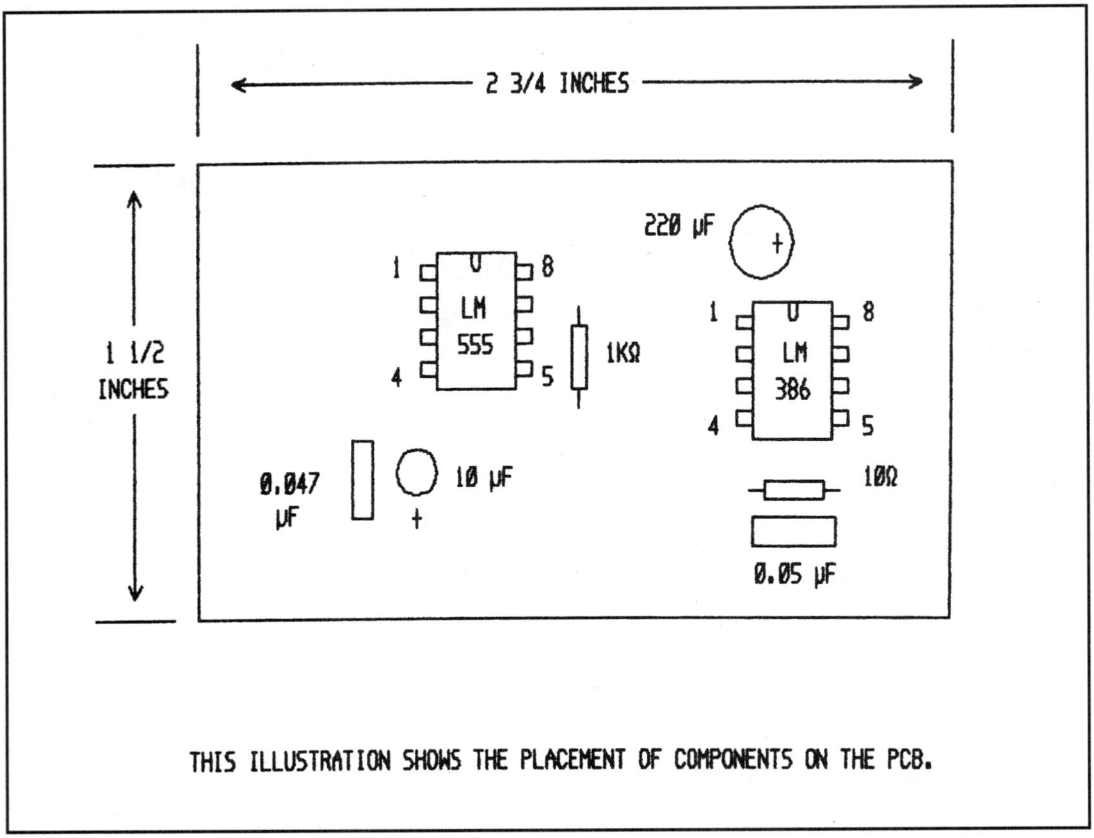

Figure 1-8. Audio generator/amp parts placement.

cut out the rough shape. For the slide switches, I started out by drilling two holes roughly the width and length of the slider, and went from there.

Next, using hobby files and sandpaper boards, shape and smooth each opening. This will provide a clean appearance when the paint is applied. Speaking of which, once all the holes are ready, paint the top surface a light color (I used yellow). This provides easy reading of the black dry-transfer lettering. Next, paint the sides of the frame with a darker color (I used a medium blue), being careful to mask off the top surface. Also, I painted a square section around the speaker holes with the blue paint, for designation of the speaker.

When the paint is completely dry, label the top with dry-transfer lettering. A light coat of clear enamel sprayed over the entire surface will protect the delicate labels. The frame is now ready to receive the various components that make up the lab. (Again, be sure the clear coat is completely dry.)

Using epoxy, glue everything to the top surface that needs to be secured. This included the SIP sockets, analog meter and LED. Then set the frame aside to let the glue cure, and move on to construct the various electronics sections of the lab.

Let's start with the signal generator and audio amplifier. Both fit nicely on a 2-3/4" (70 mm.) by 1-1/2" (38 mm.) piece of PCB material. *Figures 1-7* and *1-8*, respectively, show the foil solder side of the board and parts placement. Each of the electronic circuits could easily be done on perf board, but I elected to use PCB for durability and convenience. I have included the trace pattern for each section.

With the board in hand, the first step is to install the IC sockets, then the discrete components. Connect lead wires to the board and solder the loose ends to the various switches and potentiometers. Since all of these parts mount from the backside of the case (frame), hooking them up at this point is not a problem.

If you want to run a quick test, connect the power (5 volts for the signal generator and 12 volts for the amplifier) and a speaker to the desired output, then give it a try. The signal generator should produce

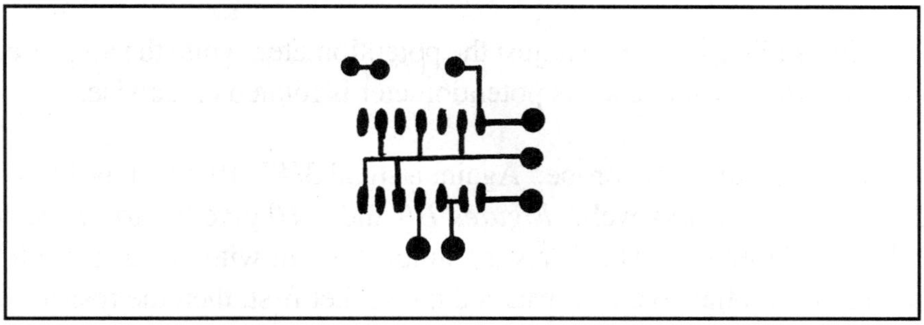

Figure 1-9. Logic probe foil pattern.

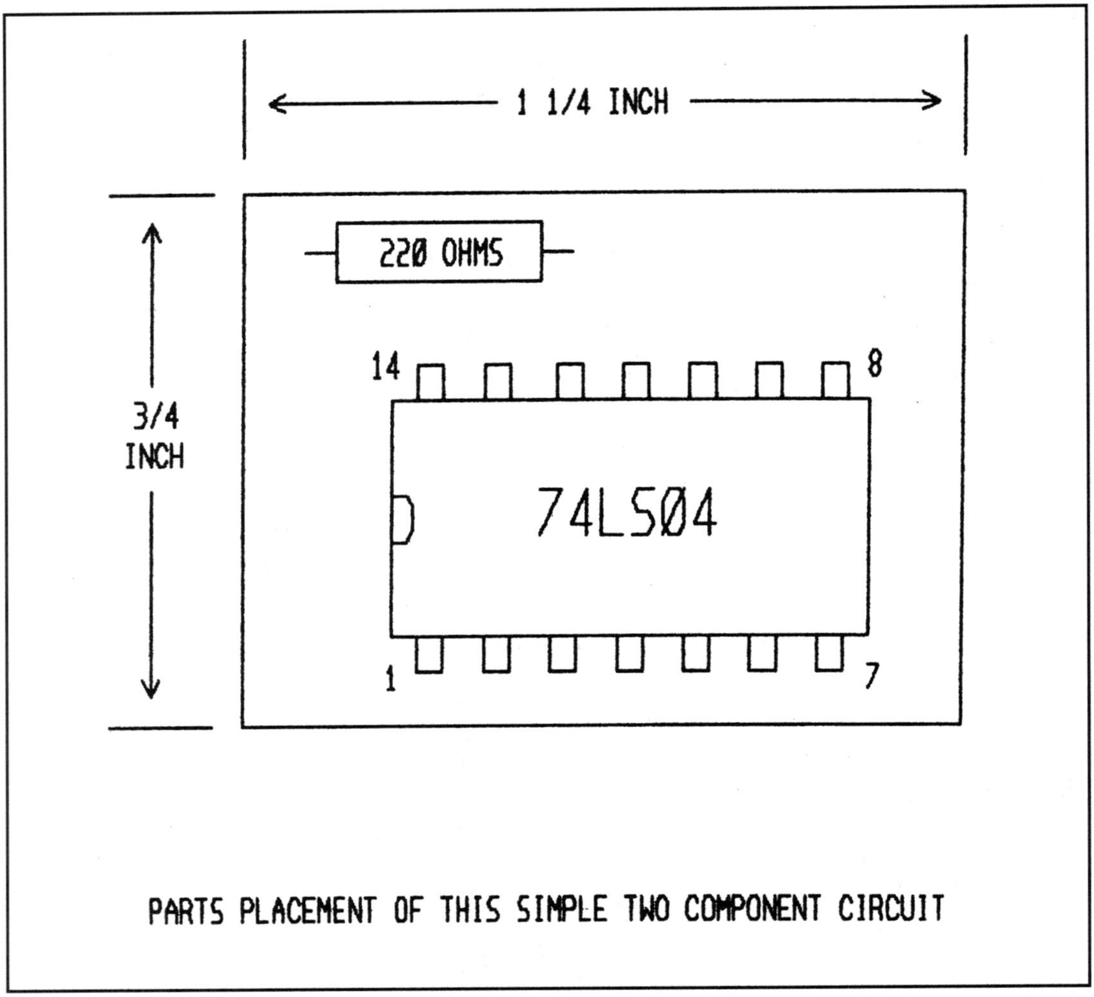

Figure 1-10. Logic probe parts placement.

a varying audio tone as you adjust the potentiometer, while the amplifier will increase in volume as its potentiometer is rotated clockwise.

Next, let's get the logic probe. Again, a small 3/4" (19 mm.) by 1-1/4" (32 mm.) PCB works well. *Figures 1-9* and *1-10* give the foil pattern and parts placement. This is a very simple circuit, with just a single IC and resistor on the board. I installed the socket first, then the resistor.

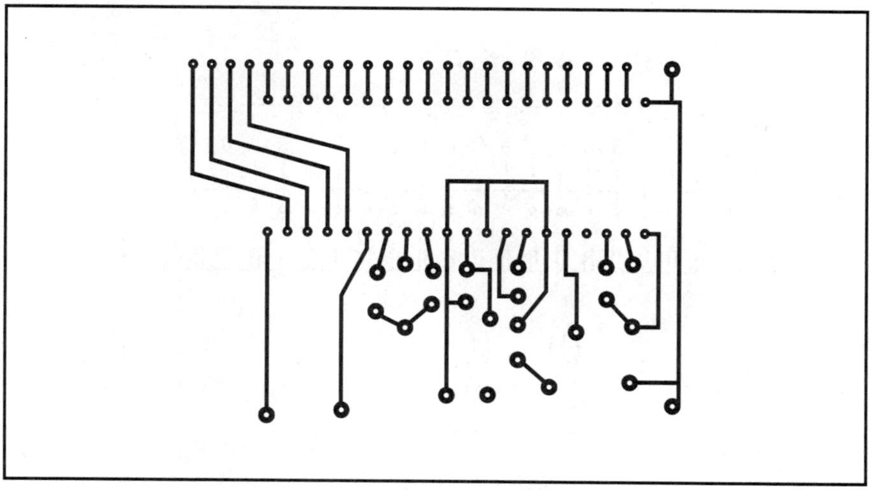

Figure 1-11. Digital voltmeter foil pattern.

Connect the lead wires, and hook up the switch and LEDs. The banana jack for the probe must wait, as it has to be installed from the front side of the frame.

A quick test here should display a lit red (high) LED when power is applied. This is because the circuit sees a floating input as a high. Touch the probe wire to ground, and the green (low) LED will illuminate.

The digital voltmeter is next. *Figures 1-11* and *1-12* show the PCB pattern and parts placement, and as usual, the IC sockets are positioned first; one on the 2-3/4" (70 mm.) by 2" (51 mm.) main PCB, and the other on the 2-1/2" (63 mm.) by 1" (25 mm.) display board. Next come the discrete components, with the hookup wires completing the initial construction.

The 7107 addresses each segment separately, as opposed to a multiplexed arrangement, so take extra care in connecting the display to the main board. I used multicolored ribbon wire for this purpose, which really helps identify the leads. *Figure 1-15* shows the pintouts for the ICC7107.

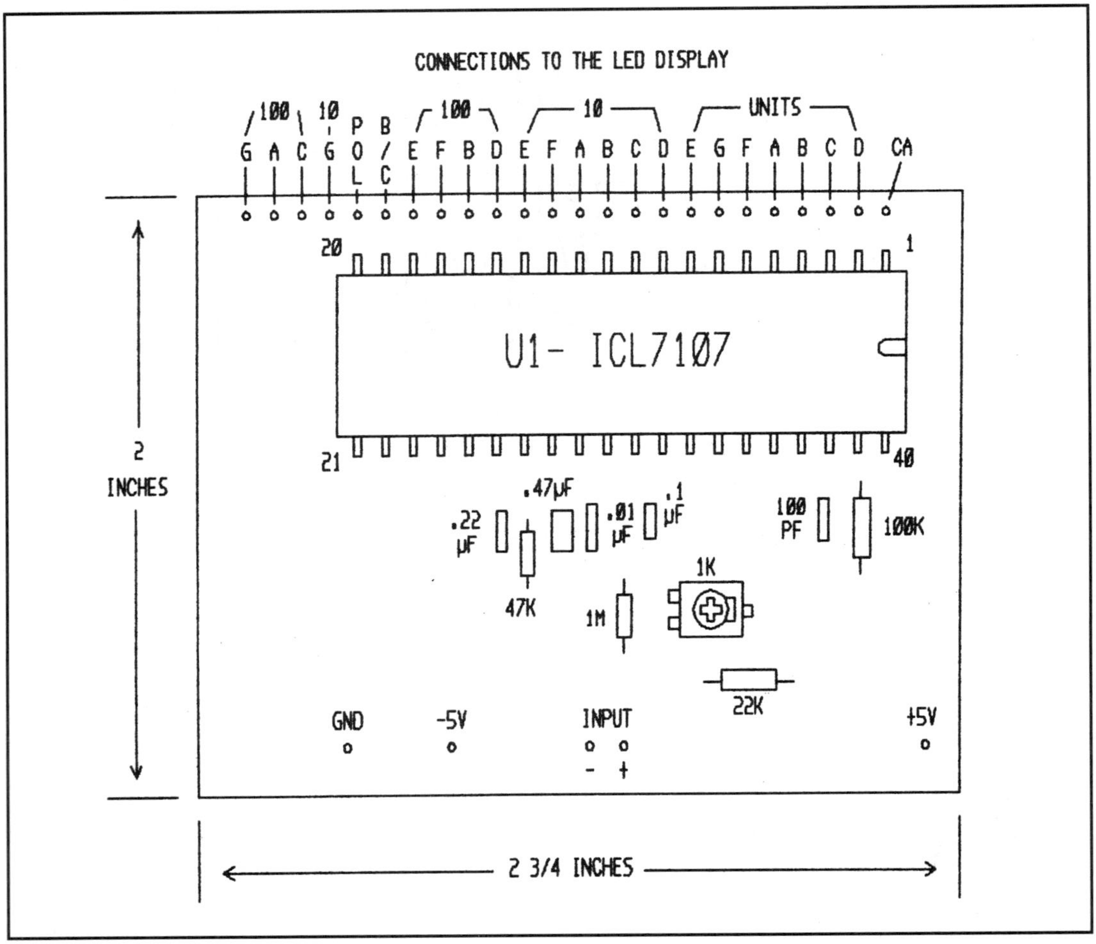

Figure 1-12. Digital voltmeter parts placement.

When everything is hooked up, apply the power. You should see a display of *000* if you have the input leads connected together. If not, the display will read *1—*. If these are your results, move on to the next step, which involves wiring the switches, with the voltage dividers soldered directly to them, to the meter's input.

Now, with the 200 mV/20V switch set at 20 volts, you can apply a known voltage to the input. By adjusting PCB potentiometer R3 until the display matches the value of the known voltage, you will have successfully (and accurately) calibrated the meter.

The input banana jacks and connections to the decimal point SIPs will have to wait until you actually install the meter in the case. This will be covered in a just minute or so. *Photo 1-1* shows a closeup view of the installed meter from the front panel.

The last of the electronics involves the power supply. A 4" (102 mm.) by 3-1/8" (79 mm.) board acts as host to all three individual supply circuits. It furnishes plenty of room for the components, including the heat sinks.

Employing *Figures 1-13* and *1-14*, first install all jumpers, then the small discrete stuff. Next will be the larger capacitors, rectifiers and finally the regulator/heat sink arrangements.

You'll note that copper has been left on both sides of the PCB to aid in heat sinking. Be sure to use a heat sink compound between the sink and the copper area, and attach the regulators securely with small bolts. This will prevent the ICs from overheating and shutting down.

Photo 1-1. Closeup view of the digital voltmeter from the front side.

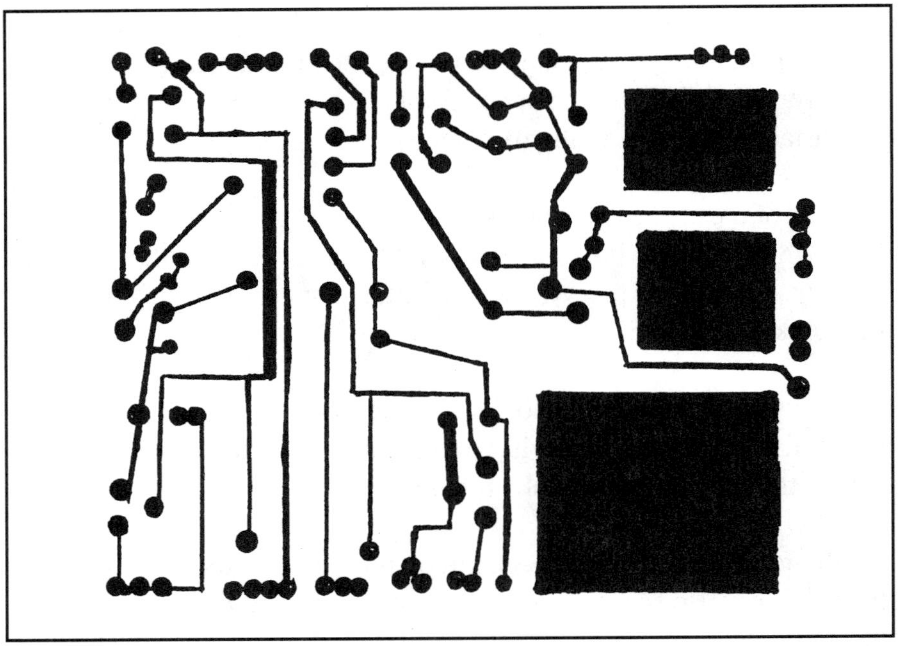

Figure 1-13. Power supply foil pattern.

It is best, at this point, to temporarily connect the transformer and confirm proper operation. In the case of the prototype, the 29-volt secondary goes to the adjustable supply, the 18-volt winding powers the +/- 12-volt section, and the 10 volts handles the +/- 5-volt system.

Once connected, power up the supplies and check each of the PCB pads for the appropriate voltage. If all is well, the lead wires can be soldered to the board, which is then ready for permanent installation. *Photo 1-2* shows the system in the case.

As for the AC access, this is handled by two banana jacks. Located on the front panel just below the six-terminal barrier strip, the jacks are connected directly to the 29-volt transformer secondary. This provides an *underload* source of around 24 volts of alternating current.

This completes the electronic circuits, and it's now time to position all this stuff on the case/frame. By now, all the paint and glue should be well cured.

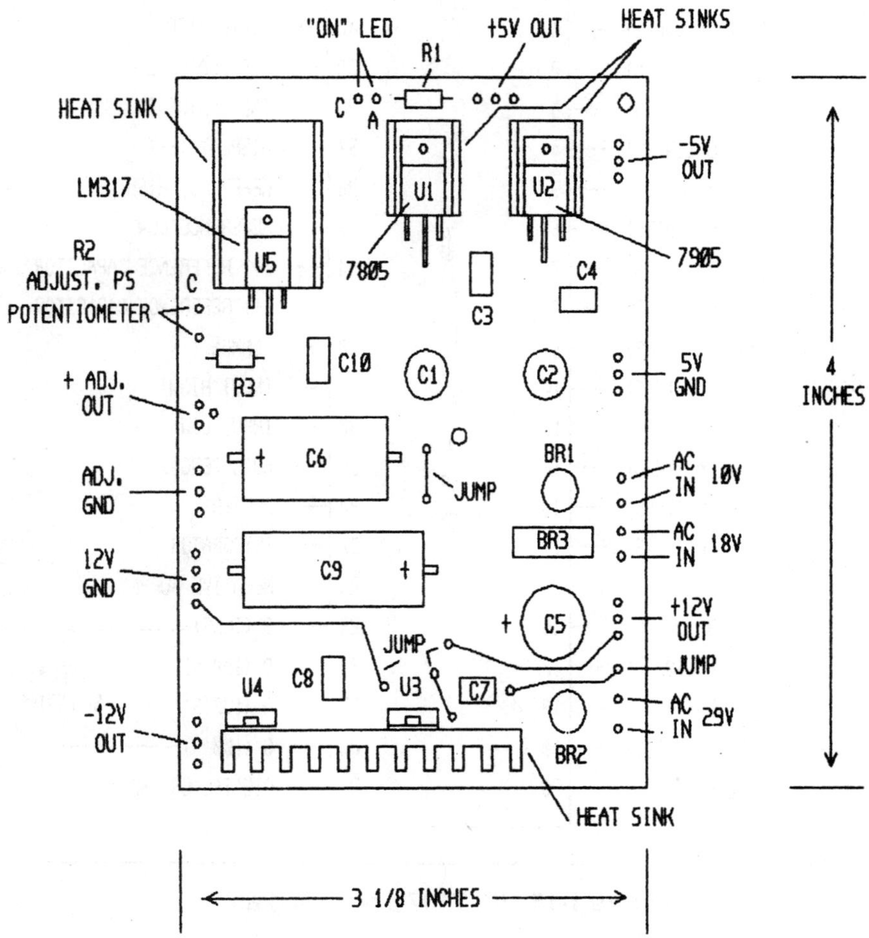

Figure 1-14. Power supply parts placement.

I began by mounting the potentiometers, switches, jacks, et cetera, on the frame. Then I came back and, using double-sided tape, placed the logic probe and signal generator/amplifier PCBs in their respective locations. For the digital voltmeter, you should first connect the three decimal point leads from the display board to the proper SIP pins, then run a line, in series with a 470 ohm resistor, to the ground pad on the main board.

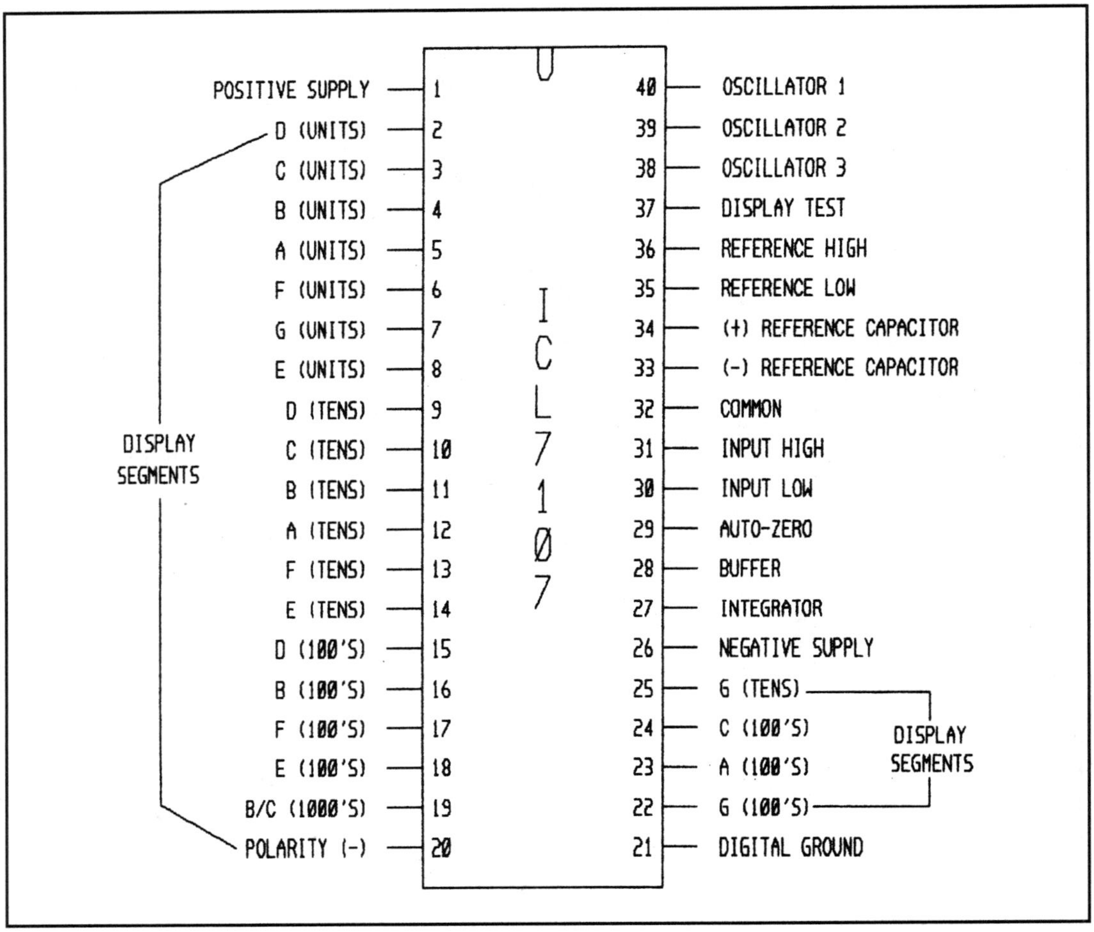

Figure 1-15. ICL7107 pinout diagram.

Now, the display board can be secured with bolts and spacers, while a single bolt and spacer holds the main board in place over and above (below) the display. This provides a compact arrangement in the upper right corner of the lab.

The power supply board and transformer are both attached with bolts and standoffs toward the center of the frame area. As seen in *Photo 1-3*, this makes for an orderly arrangement.

Next, it is time to make the interconnections between the SIP sockets, electronics and other components on the frame. This can be a little te-

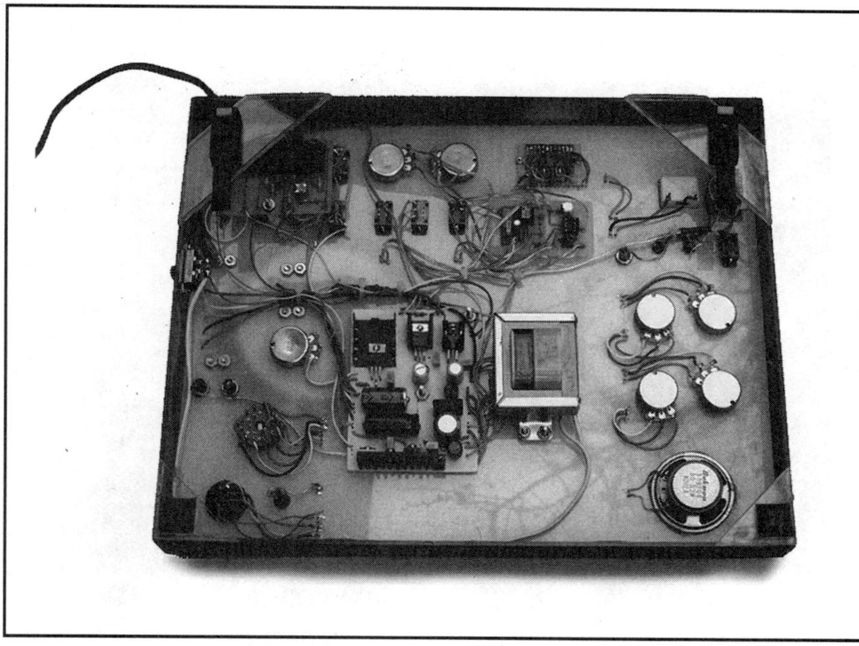

Photo 1-2. The prototype lab system in its case (back side).

dious, especially with the SIPs, as their pins are only 0.1 inch apart; watch carefully for solder bridges.

For the power supply access, drill small holes next to the barrier strips for lead wires. Then attach the cables to the appropriate terminals on the strips. The line cord comes in from the back upper right corner, and runs to the *power-on* switch located on the right side. Trail a length of zip cord to the power transformer's primary winding terminals and secure it in place with epoxy.

The only other tricky device to install is the interchangeable seven-segment LED display socket. Actually, it isn't very tricky, but you have to be careful to match the display's leads to the markings on the top surface of the frame.

I marked my case in alphabetical order (*a* through *g*), but this is not the way the display is configured. For example, the top row of pins

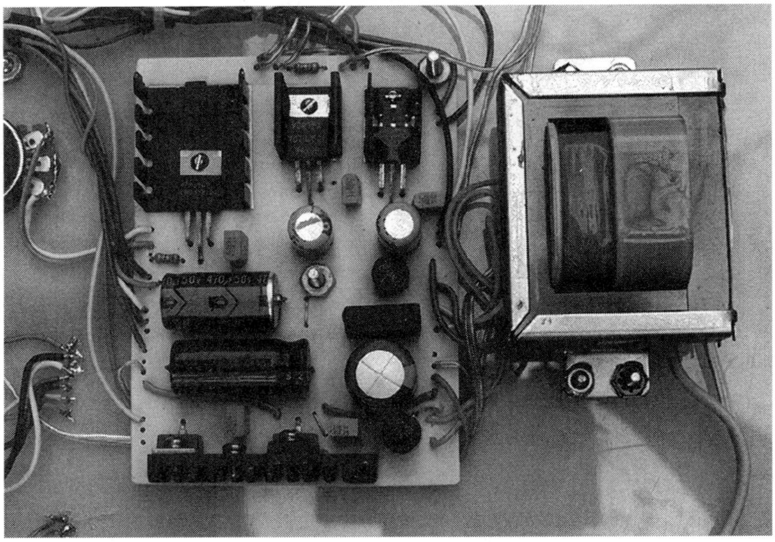

Photo 1-3. A closeup view of the power supply and transformer.

are designated b1, a1, f1, CA, CA, b2, a2, g2 and f2. Thus, the backside wiring has to reflect the pin designations.

I saved the solderless breadboards for last. Using the adhesive backing on the units, I put them in the space left on the top surface (bottom center and up). This position provides easy access to everything else in the lab.

To complete the construction, I epoxied the speaker in place from the back side; but you can use a speaker bracket if you have one. Also, the transformer is a little taller than the frame is deep, so I cut some 1/4"-thick plexiglas pieces to fit across each corner. These have to be glued in position as well; PVC cement works just fine for this.

With the corner pieces affixed, install rubber feet to provide stability and additional height. With these in place, the frame sits evenly on the bench with ample clearance for the transformer.

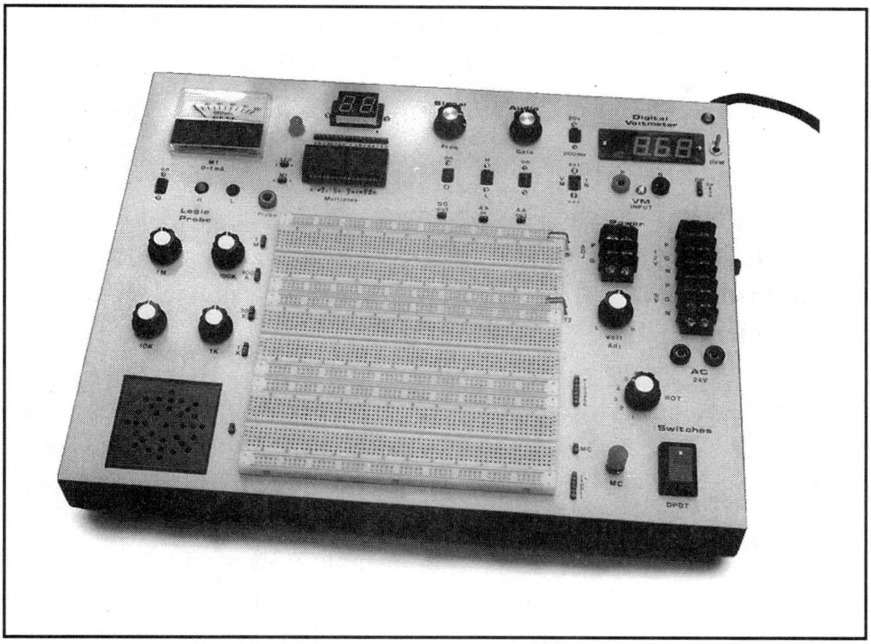

Photo 1-4. Complete prototype station.

Attach the knobs to the potentiometers, put the rotary switch on top, and she's-a-done (see *Photo 1-4*). The next step is to try everything out to be sure it works.

Testing and Application

Testing is a reasonably simple matter of checking each component or circuit to make sure it does what it is supposed to. A good place to start is with the power supply.

Using a VOM (volt-ohmmeter) or DMM, test the barrier strip terminals for proper voltages and polarities. The adjustable supply should, of course, show a decreasing or increasing voltage as the potentiometer is moved back and forth.

Each of the switches should display proper operation when checked at their respective SIP sockets, as should the potentiometers. The analog meter can be tested with a 1-1/2 volt AA battery, which will normally bring about 1/2 full scale. The LED will glow when 3 volts are applied to the proper terminals.

For the digital voltmeter, set the voltage switch to *20V* and the input switch to *ext* (external), and hit the power. The display should read *000*. To set a decimal point, move a jumper from the *DP* SIP socket to the desired position (2, 3 or 4). This will display a decimal point at the location selected.

Now, set the voltage switch to *200 mV*, and the display will show something like *002* or *003*. This is due to the fact that, at this sensitivity, the meter picks up some of the stray voltage floating around.

Next, set the input switch to *aps* (adjustable power supply). As you advance the adjustable power supply's potentiometer, the display will indicate the output. If you haven't already calibrated the meter, this is a good time to do it. Measure the barrier strip output voltage, and adjust the voltmeter's potentiometer (R3) to match the reading. Simple, eh?

To test the signal generator and amplifier, run leads from the output SIPs of whichever device you're testing to the speaker SIPs. The signal generator will produce the audio tones while the amplifier should exhibit a low-level hiss. Add an input source and the amp will, well, amplify it. Either of these tests also confirms proper speaker operation.

The logic probe should function as described before; that is, on power up, the high LED will light, and touching the probe to a ground level will illuminate the low LED. Also, touching the probe to a high level will not change the LED status, as the high LED is already lit. Consequently, any point within a circuit being tested that does not change the LEDs is high.

Regarding the seven-segment displays, it's is just a matter of testing each segment to make sure it illuminates. Using a 3-volt DC source, touch the positive lead to the common anode SIP, then touch each of the segment terminals. As you go along, the *a*, *b*, *c*, et cetera, lights will light, verifying the display. For common cathode units, use the same technique, just reverse the polarity.

If everything checks out, the lab is ready for service. At this point I hope that the intended service of the lab is obvious. If not, I have failed in my mission here. However, in case I have, let's take a quick look at how this lab will be of value to you.

Much of this will be a repetition of information from the introduction and elsewhere in the book. This reminds me of a Will Rogers story. It seems Will was once asked to comment on the profession of someone responsible for a certain deed, and his reply was, "It must have been a politician, or a crook. Whoops, there I've gone and repeated myself."

Anyway, as stated earlier, the lab is designed to make your prototyping efforts less painful. Providing easy access to some of the more problematic components, (problematic in terms of breadboard application) helps keep your projects manageable.

The addition of the logic probe, signal generator, audio amplifier and digital voltmeter furnishes some of the most frequently-used prototyping tools. Also, the voltmeter and amplifier can frequently end up as part of a design. Their presence here, in ready-to-use form, can speed up the development process.

The spacious 2,490 tie point solderless breadboard will accommodate virtually any project your heart desires, within reason. I wouldn't try designing a 200-megahertz Pentium-controlled personal computer on this system. Well, maybe parts of it, but not the whole thing. However, endeavors as complex as a microprocessor trainer, communications transceiver or frequency counter shouldn't present any problems. Thus, your prototyping latitude is bound by little more than your creativity.

On a last note, an additional plus to this system is size. At 11" x 14", it isn't going to steal half your workbench, unless you have an awfully small workbench. While the layout might not seem overly important at first glance, it will well prove its worth down the line. Prototype labs have a mysterious ability to expand before you even notice.

Conclusion

With your trusty prototype lab in hand, you are now ready to tackle the most ambitious projects you can conjure up. It will also provide a quick and handy way to evaluate that WWV receiver circuit you saw in an electronics magazine. Yes sir, you've really got a jump on that bad boy. Realistically, though, the lab is a *must have* for any electronics workbench. As serious hobbyists, we all need the convenience the lab provides. Not only does it speed up the work, but it also allows your mind to concentrate on the project instead of worrying about how to keep all those wires connected.

So, break out the proverbial junk box and put this beauty together. I can almost guarantee you won't regret it, and you may well be labelled a genius or visionary by your peers. Whoops, there I've gone and repeated myself again.

Chapter 2
A Multi-Output Test Bench Power Supply

Introduction

A must-have for any electronics lab is a dependable direct-current (DC) power source. There are a number of ways to obtain this necessity, such as purchasing a commercial unit, buying one from surplus computer suppliers, or (of course) creating your own.

The problem with most commercial units and computer systems is that often you don't get exactly what you want or need. Dedicated workbench supplies are usually overpowered (in terms of amperage) and expensive, while computer units were designed for a particular machine. This can mean unneeded voltage outputs that are of use only to a computer. Hence, the best approach is to build what you need.

I know what you're saying: "Yeah, he's writing this book on how to build your own equipment, so he would say that." Okay, there is a degree of truth in that; but by constructing the system yourself, you can ensure that it will have the basics in addition to the whistles and bells YOU want, not what manufacturer XYZ wants you to have.

This is not meant as an indictment of commercial systems. A lot of time, reflection and work goes into developing the commercial units, and I think the companies truly feel they are producing the perfect product. It's just that I have yet to find one that completely filled my requirements. Invariably, to get the outputs I want, I have to buy a bunch of stuff I don't want.

BENCH POWER SUPPLY PARTS LIST

SEMICONDUCTORS

U1	LM317T Positive Adjustable Regulator
U2	LM337T Negative Adjustable Regulator
U3	7805T 5 Volt Positive Regulator
U4	7905T 5 Volt Negative Regulator
U5	7812T 12 Volt Positive Regulator
U6	ICL7107 3 1/2 Digit A/D Converter
LED1-5	Jumbo T-1 3/4 Light Emitting Diodes (Various Colors)
BR1-3	Bridge Rectifiers (W04M, or Equivalent)
DSP1	2- MAN6710 Common Anode Dual Digit Displays (A number of other displays will also work)

RESISTORS

R1, 4	5,000 Ohm Panel Potentiometers
R2, 5, 7, 8	220 Ohm 1/4 Watt Resistors
R3, 6, 9	1,000 Ohm 1/4 Watt Resistors
R10	470,000 Ohm 1/4 Watt Resistor
R11, 16	1,000,000 Ohm 1/4 Watt Resistors
R12	25,000 Ohm PCB Potentiometer
R13	22,000 Ohm 1/4 Watt Resistor
R14, 15	100,000 Ohm 1/4 Watt Resistor
R17	470 Ohm 1/4 Watt Resistor

CAPACITORS

C1, 2, 9	1,000 Microfarad/63 Volt Electrolytic Capacitors
C3, 4, 10	4.7 Microfarad/50 Volt Tantalum Capacitors
C5, 6	470 Microfarad/25 Volt Electrolytic Capacitors
C7, 8	1 Microfarad/50 Volt Tantalum Capacitors
C11	O.22 Microfarad Mylar Capacitor
C12	0.047 Microfarad Mylar Capacitor

C13	0.01 Microfarad Mylar Capacitor
C14	0.1 Microfarad Monolythic Capacitor
C15	100 Picofarad Mica or Ceramic Disk Capacitor

OTHER COMPONENTS

J1-8	Red and Black Banana Jacks
S1	SPST Pushbutton Panel Switch
S2	2P6T Rotary Switch
F1	1 Amp Fuse

MISCELLANEOUS

Knobs, LED Holders, Fuse Holder, Clear Red Plastic Display Lens, Solder, PCB Materials, 8" x 6 1/2" x 2 1/2" Instrument Case, Paint, Dry Transfer Lettering, Hardware, Hook-Up Wire, Heavy Duty Heat Sinks, Line Cord and Plug, 40-Pin IC Socket, Etc.

So, let us take a look at how to develop what is, in my opinion, the perfect bench power supply (almost perfect, anyway). YOU, however, might disagree with me, and therein lies the beauty of this project. This power supply can be manipulated, twisted or modified in virtually any fashion that suits you.

Theory

The hub of this system is our old friend, the linear regulator. As we saw in the last chapter, linear regulators are simple to use and provide rock-solid voltages. Additionally, the *T* versions furnish up to 1.5 amps of current, and there is considerably less noise in the output than with the switching variety.

With that said, let's set up some parameters for our system. It has been my experience that an amp-and-a-half is more than sufficient current for a vast majority of the projects I build, so the *T* series is perfect.

Our next consideration is the outputs. Again, I base this on past practice, but positive and negative (+/-) 5 volts, positive 12 volts, and positive and negative adjustable voltages have always filled the bill. The +/ - 5 volts allows for the operation of logic and computer-related circuits. The positive 12 suits linear systems, and the variable supplies handle everything in between.

The last item is not essential, but I like it anyway; that is, a digital voltage readout. A standard analog meter would also work, or you could leave this off altogether; however, I think you will enjoy this feature and the digital display precision. Using a voltmeter keeps you constantly informed of the supplies' output. While this is not terribly important with the fixed 5- and 12-volt legs, it is most convenient with adjustable supplies. The digital voltage readout can also be configured to read output current. More will be said about that later.

So, there are our demarcations. When completed, the power supply will furnish essentially all of your needs. Additionally, the design allows you to use any or all of the outputs simultaneously, since each is the product of an independent circuit employing a separate secondary of the power transformer.

A quick sidebar here regarding the transformer. This circuit requires a transformer with three secondary windings, and this might at first glance seem problematic. Actually, these are not as hard to come by as you might think.

For the prototype, I employed an inductor purchased from a surplus company (see parts list); however, by the time you read this, the transformer may no longer be available. I have listed several distributors that carry a substantial line of transformers, and chances are one of them can help you.

Another wealth of such units is old, broken or otherwise nonfunctional video recorders, audio equipment and some test equipment (vacuum tube testers, for example). So, if the transformer becomes a problem, look around. You will probably find one right under your nose.

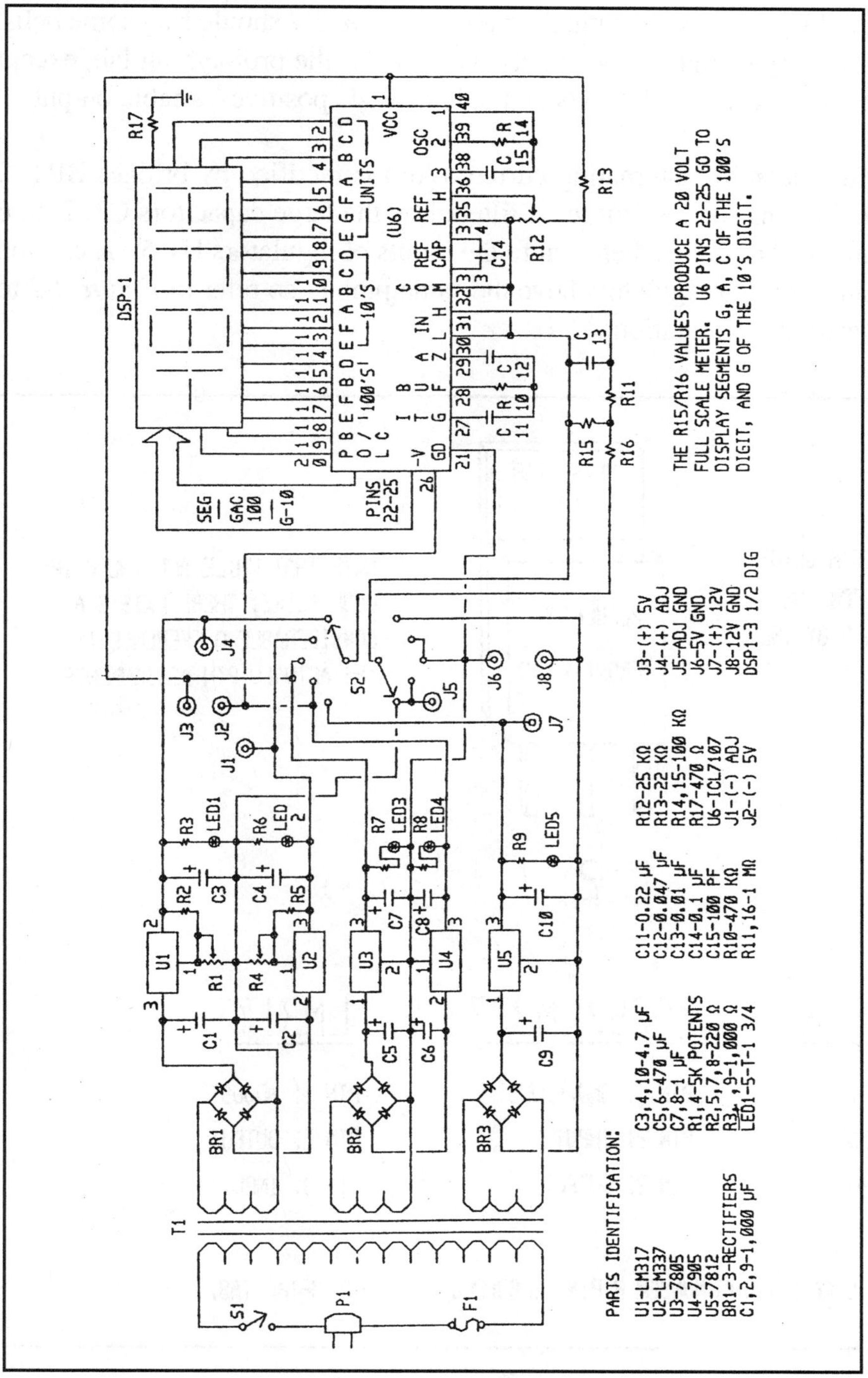

Figure 2-1. Multi-output power supply schematic diagram.

Looking at the schematic diagram in *Figure 2-1* should ring some bells. It is very similar to the supply we built for the prototyping lab, except there we used +/- 12 volts and only a single positive-variable output.

The incoming alternating current (AC) is rectified by bridges BR1, 2 and 3, and the DC output is filtered by the large capacitors C1, 2, 5, 6 and 9. The DC is then sent to the inputs of regulators U1-5. A caution here: not all regulators have the same pinout, so refer to *Figure 2-2* to verify the orientation.

THE FOLLOWING IS A GUIDE TO THE PINOUTS FOR THE 5 REGULATORS USED BY THE POWER SUPPLY.

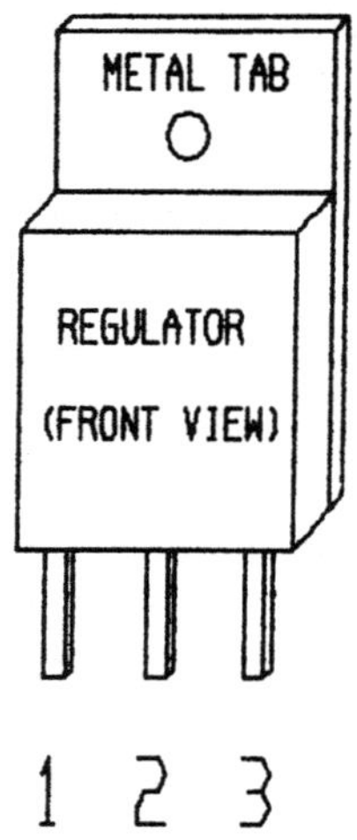

NOTE THAT WHILE ALL THE CHIPS LOOK ALIKE, THERE EXISTS A CONSIDERABLE DIFFERENCE IN THE ACTUALL PIN ASSIGNMENTS.

7805/7812	7905/LM337	LM317
PIN 1: INPUT	PIN 1: GND/ADJUST	PIN 1: ADJUST
PIN 2: GROUND	PIN 2: INPUT	PIN 2: OUTPUT
PIN 3: OUTPUT	PIN 3: OUTPUT	PIN 3: INPUT

NOTE: IN ALL CASES, THE NUMBER 2 PIN IS CONNECTED TO THE METAL TAB.

Figure 2-2. Regulator pinout diagram.

You'll note that in the case of both the +/- 5 volt and the +/- adjustable supplies, a *hot ground* arrangement is used. That is, the ground is derived from one side of the AC secondary winding. If you use a transformer with center taps, these will accommodate the ground lines quite nicely. However, bear in mind that the center-tapped version will only produce half the full voltage of the winding. Either approach will work, so take your choice.

Next, notice that the positive side of the bridge goes to the positive regulator input, while the negative side feeds the negative regulator. Also, observe the polarity of the capacitors. It is normal for the positive rails of all sections; but for the negative rails, the *plus* leads are applied to ground.

The same principle applies for the light-emitting diode (LED) output indicators. All is normal for positive, but for the *minus* side, anodes are connected between ground and the negative line. Capacitors C3, 4, 7, 8 and 10 are the output filters. Actually, these serve three functions. The first is to filter any remaining transients on the lines. The second is to prevent the regulators from oscillating. The last is as a *bypass* to furnish short-term power storage when a system needs it.

The last note of importance involves the divider networks incorporated in the variable supplies. One leg of each is the panel-mounted potentiometer (R1 or R4) that allows for voltage adjustment. The second leg, fixed resistors R3 and R6, feed back to the positive or negative output lines. This is the normal arrangement for the 7800/7900 linear adjustable regulators.

Switch S1 provides ON/OFF operation, and fuse F1 (1 amp) protects the circuit from voltage surges, overloads and internal failures. Such internal problems might include shorted components or regulator disasters. In the event of trouble, the fuse will blow and prevent total destruction of the supply. This is always good news. The origin of the difficulty, on the other hand, can be bad news; but at least the system has a built-in safeguard against really serious damage and danger to the user.

So, there you have the main section of this project; the linear power supplies. Now let's have a gander at the digital voltmeter that monitors the outputs. This segment utilizes one of the best chips ever developed, at least as far as I'm concerned. Originally built by General Electric's Intersil division, it is now produced and distributed by Harris Semiconductors. The chip is the ICL7107.

The ICL7107 houses in a single 40-pin dual in-line pin (DIP) package a complete 200-millivolt full-scale digital voltmeter. This includes the analog-to-digital (A/D) converter, clock, reference and output decoders, and drivers capable of illuminating instrument-size (0.56 inch) LED seven-segment displays. Its cousin, the ICL7106, addresses liquid crystal displays (LCDs).

For many years, the ICL7107 chip has been the heart of most digital multimeters, since any quantity that is a voltage, or can be converted into a voltage, can be measured and digitally displayed by this device. Furthermore, voltage dividers can expand the chip's range to as high as 2,000 volts. Very handy, indeed.

As with many analog-to-digital converters, a *dual slope* technique is employed here. The initial slope is that of integrating the input high and input low differential for a fixed period of time.

The second slope is the *de-integration phase*. Here, input low is sent to analog common (pin 32), and input high is connected across a previously-charged reference capacitor. The capacitor brings the output back to zero, with the input voltage represented by the time it takes to do so. The remaining circuitry then decodes the result and displays it. Simple, eh? Yes, but most effective.

In the last few years, the price of the 7107 has dropped dramatically (single units can now be purchased for around $3), making the chip very attractive for many projects. As I commented in Chapter 1, this is still one of the best ways to measure and display DC voltage.

Another nice feature of the 7107 is *auto-zeroing*. This ensures a zero reading for a zero input, accomplished by an external network consisting of C11, C12 and R10. Capacitor C14 is the reference capacitor previously mentioned, while potentiometer R12 and resistor R13 control the reference voltage.

C15 and R14, connected to pins 38 and 39 respectively, work in conjunction with pin 40 to set the oscillation frequency. This controls the sampling rate, or the number of times per second a new measurement is taken. For the recommended values, a 48-kilohertz clock is established, providing three readings each second. This frequency can also be crystal or external oscillator controlled.

One drawback to the 7107 is that it requires a split (+/-5 volt) power supply. However, for our unit, this is easily furnished by the main board's 5-volt section. A separate positive/negative 5-volt supply can be included if desired; but for space consideration, I used what I already had. Since the 7805/7905 arrangement can produce up to 1.5 amps on each rail, current drawn by the meter is practically negligible.

Rotary switch S2 allows any one of the five outputs to be applied to the voltmeter's inputs. In this fashion, you can quickly check the outputs, both underload and free-running.

Since I configured the basic 7107 system to read 2 volts full scale, a voltage divider is needed to condition the higher levels that the power supply will produce. For our unit, resistors R15 and R16 do the honors. With values of 100 kilohms and 1 megohm, respectively, a ratio of 10 is established. This allows the basic meter to measure voltages as high as 20 volts with the resolution to one-hundredth of a volt (0.01).

Depending on the power transformer used, the adjustable supplies may provide more than 20 volts, as the regulators will accommodate up to 37 volts. This can be handled in one of two ways. First, you can use a separate digital multimeter (DVM) to check outputs that send the internal meter to overscale/overflow (a reading of *1—*). The second way would be to make R15 10 kilohms instead of 100K. This establishes a ratio of 100, resulting in a full-scale reading of 200 volts. This will definitely cover anything the power supply is capable of producing, but it also lowers the resolution to one tenth of a volt (0.1).

On a final note, I mentioned earlier that the meter could be used to measure amperage if desired. Personally, I haven't found this feature useful, but if you want to know the total current a prototype is using, for example, then it will be helpful. Anyway, this is done by removing the voltage divider and replacing it with a *shunt* resistor. What is a shunt resistor, you ask? It is a high-wattage unit of low value that is connected across the voltmeter's inputs.

Figure 2-3 illustrates this procedure. The recommended resistor value is 0.1 ohms at 5 watts. This will produce a 2-amp reading at full scale, which should cover anything you work with. Remember, the supplies can only provide 1.5 amps.

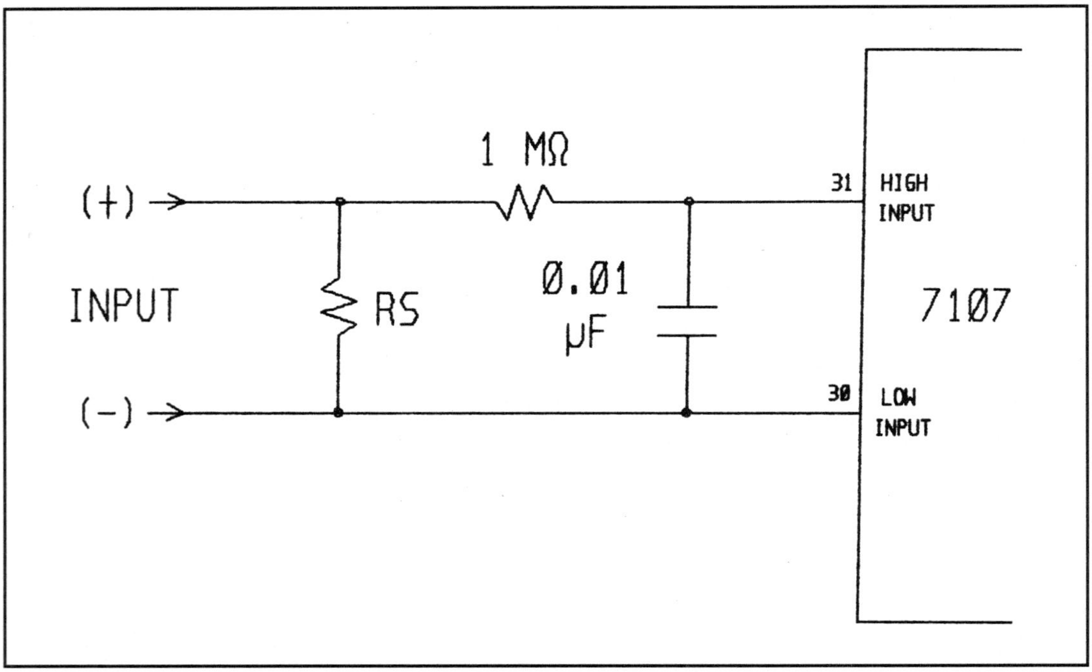

Figure 2-3. Amperage readings using a shunt resistor.

A good deal of power dissipation occurs across the resistor; hence the large, 5-watt rating. Actually, a 2- or 3-watt component would do as well, but the 5-watters are easier to find. Naturally, an additional switch will be necessary to determine which resistor configuration goes to the meter's input.

That about covers theory. Banana jacks provide the outputs, and various color LEDs indicate whether or not the various stages are func-

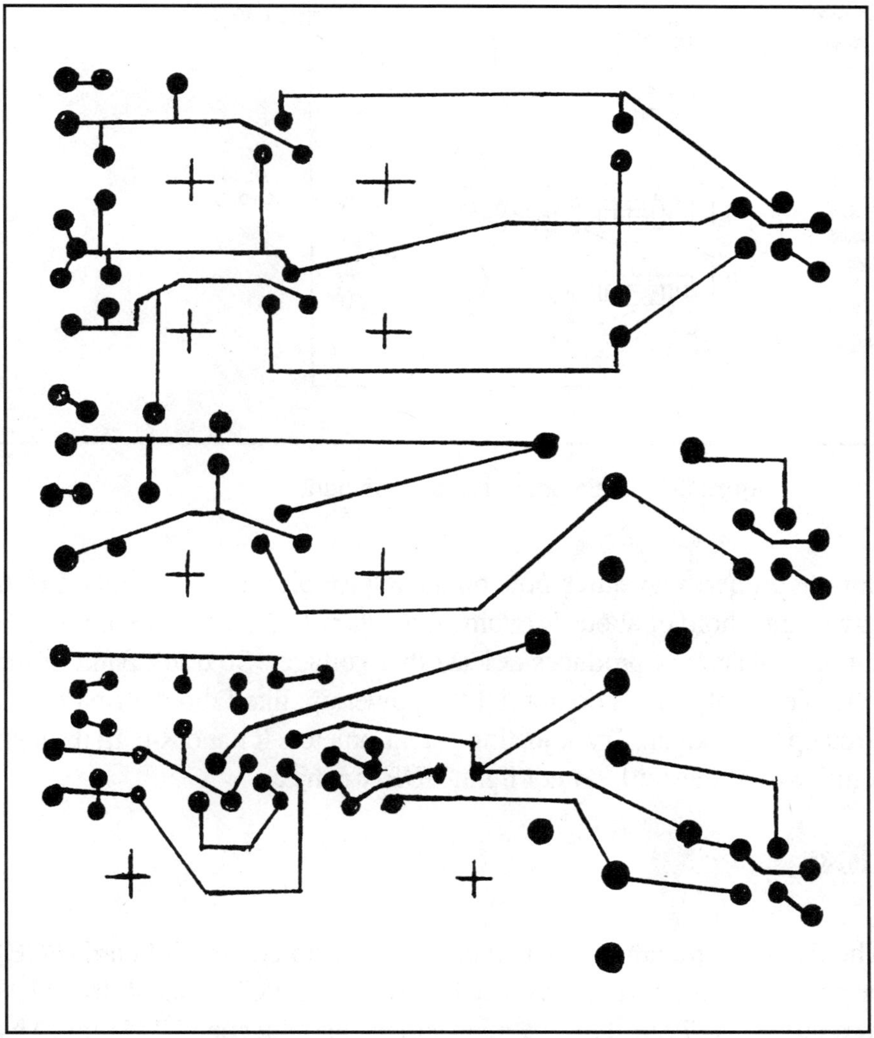

Figure 2-4. Power supply foil pattern.

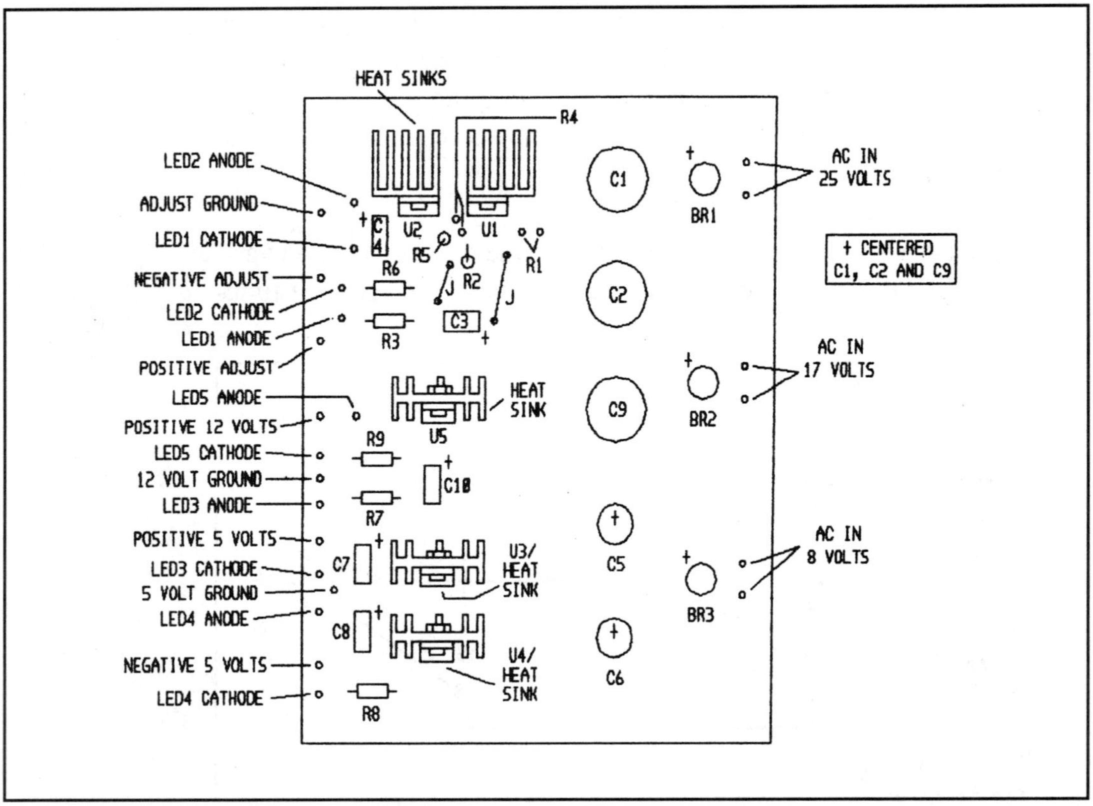

Figure 2-5. Main board parts placement.

tioning properly. A quick note on the adjustable supplies: since LEDs have a threshold of about 1.7 volts, indicators 1 and 2 will not light until supply adjustment produces at least that voltage. So don't panic if the variable supply LEDs do not light at power up, like I did the first time I fired up this gadget. Try adjusting potentiometers R1 and R4. If, through a full swing, they still do not light, THEN panic.

Construction

The first determination is whether to use a printed circuit board (PCB) or point-to-point wiring. As usual, I chose two PCBs; one 4-1/4" (108 mm.) by 5" (127 mm.) for the main supply, and a second 2" (51 mm.) by

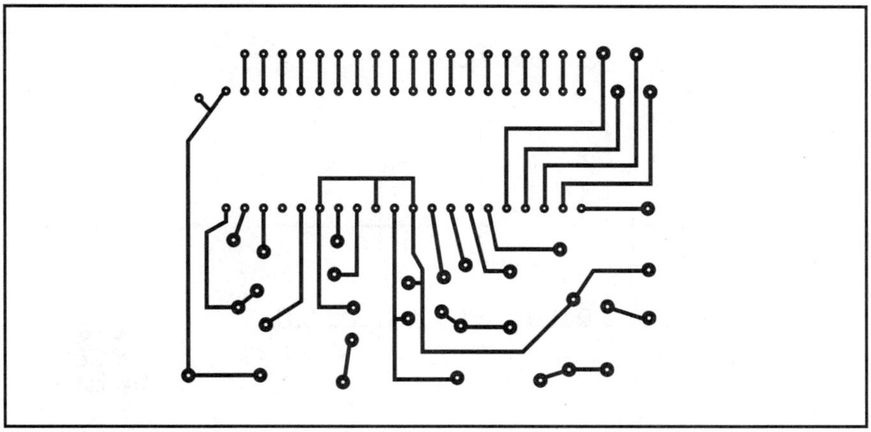

Figure 2-6. Voltmeter foil pattern.

2-5/8" (66 mm.) for the digital voltmeter. These patterns are illustrated in *Figures 2-4* and *2-6*. Parts placement is seen in *Figures 2-5* and *2-7*.

Actually, had I had my wits about me, I would have included both patterns on a single board, leaving a notch in the upper right corner for the transformer. As it turned out, I was sleeping when I designed this part of the project, thus the two separate boards. However, if you are so motivated, my oversight is easily rectified. A third 1-1/8" (29 mm.) by 2-5/8" (66 mm.) PCB handles the display, but since it has to be mounted on the front panel, it could not be part of the main board. So there! I wasn't completely asleep. Just dozing.

Anyway, for the main board, install the jumpers, resistors and small capacitors first, then the larger capacitors. Next comes the bridge rectifiers and regulators; be sure to observe proper orientation.

Also, use heavy-duty heat sinks on all regulators, as this circuit will generate some warmth, and not necessarily the nice kind. Seriously, without sufficient heat sinking, heavy loads will result in the supply shutting down. Each of these regulators has built-in thermal protection that prevents damage or destruction from excessive heat. They simply turn off until they cool down.

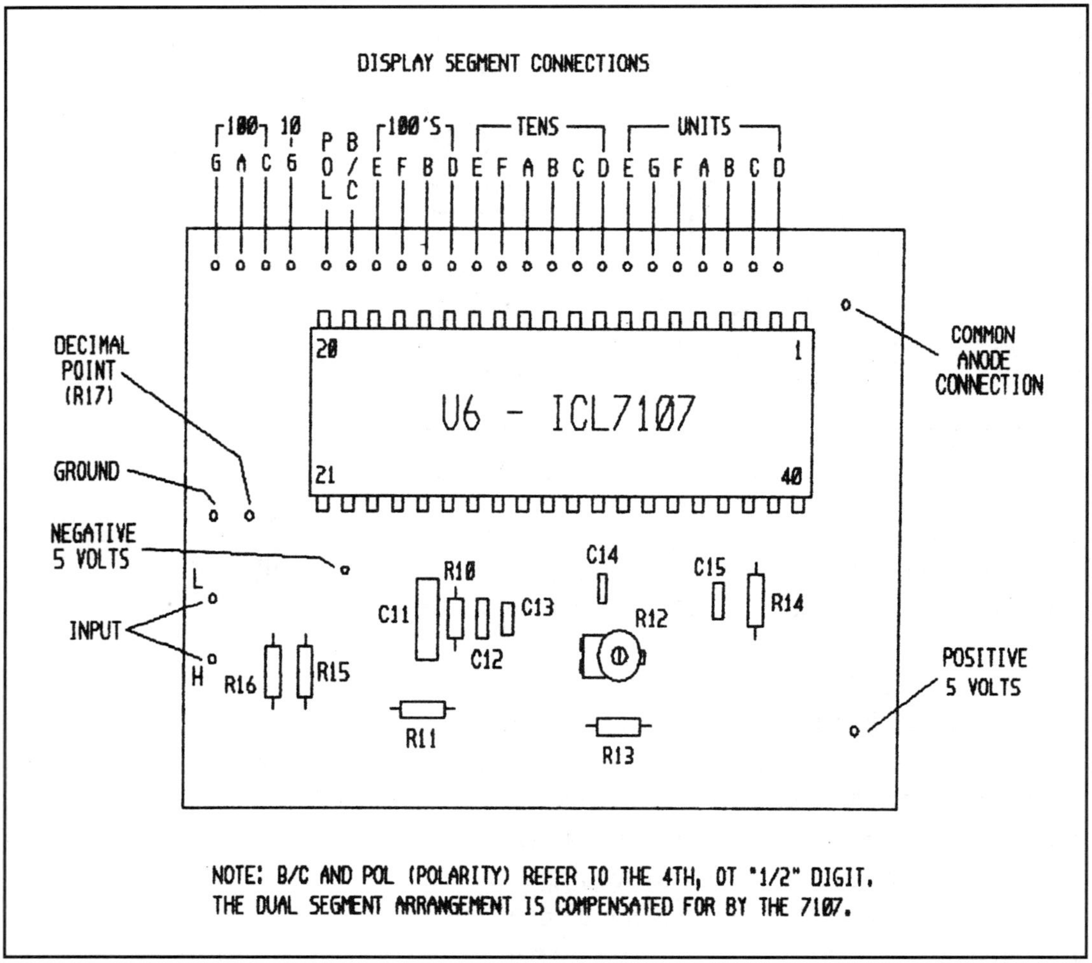

Figure 2-7. Digital voltmeter parts placement.

Once everything is on the board, connect the transformer secondary leads to the proper pads. Now solder appropriate lengths of hookup wire for the five LED indicators, two adjustment potentiometers and the three outputs. Of course, the +/- 5 volt and adjustable supplies will each require three output leads.

The next step is to connect the LEDs and potentiometers to their respective leads. Be sure the diode anodes and cathodes go to the correct wires, since they won't light if reversed.

At this point, temporarily hook a power cord to the transformer primary and run a quick output test. Any standard digital multimeter can be used to confirm proper operation. If everything checks out, move on to the voltmeter.

First comes the 40-pin IC socket, then the discrete components. Follow this with the various leads for input, power and the display. There are a slew of these, so be patient. Part of the reason for this is the fact that the 7107 addresses each display segment individually, rather than employing a multiplexing arrangement. This is both good news and bad. The bad news is that 23 leads are needed for the display alone. However, the good news is that a multiplexed display is not required.

Photo 2-1. Stuffed board for the multi-output bench power supply.

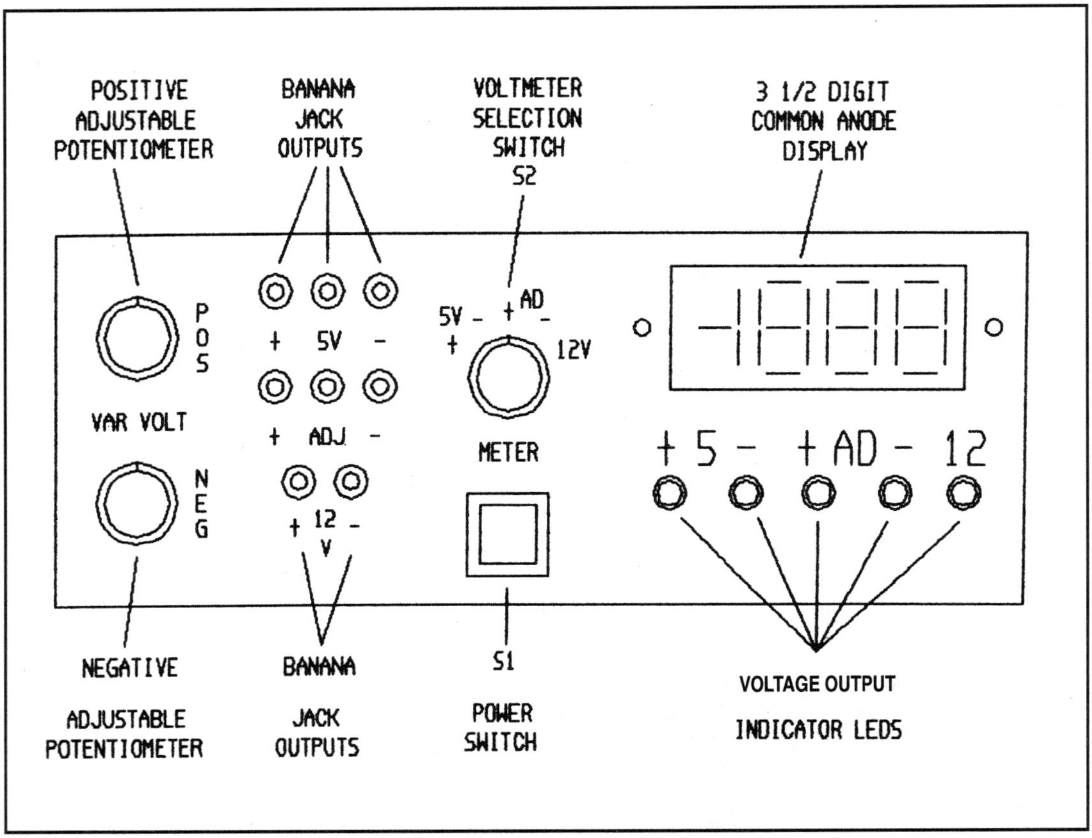

Figure 2-8. Power supply front panel template.

These displays are getting harder to find all the time, and if you have ever had to hard-wire one, you will appreciate this aspect. However, the individual segment approach does require more than double the number of lead wires.

Set the DVM board aside for the moment and turn your attention to the display board. This one is very simple, and is designed to use MAN6710 common-anode dual-digit displays. However, many similar 0.56" dual displays employ the same pinout, and can be substituted for the MANs. If you have some dual common anode units on hand, check them out. They might work. They will have to have pins running along the tops and bottoms, not down the sides.

The IC socket goes on first, then the 470-ohm decimal point resistor. Last, connect the leads from the DVM board to the display. Do this carefully, as mistakes here can result in a display that reads like hieroglyphics.

With that done, temporarily connect the main board's 5-volt supply to the meter and give it a try. You should see a display of *000*. For a reading of *1*— (overflow), try shorting the inputs to produce the three zeros. If you get anything else, disconnect the power and double-check your work.

This completes most of the wiring; now comes the case. A standard 8" (204 mm.) by 6-1/2" (166 mm.) by 2-1/2" (63 mm.) instrument case is used for the prototype. As can be seen in *Photo 2-1*, space is tight. You can use a larger case if needed or preferred.

Figure 2-8 is a template for the front panel. This layout worked well for the prototype, but can be rearranged if desired. The main point is to allow room for all the jacks, switches, LEDs and display.

The round openings are done with standard drill bits. The display aperture starts with small round holes at each corner; then, using a small hobby-style saw, is cut out by connecting those holes. A variation involves a sharp hobby knife and metal straightedge to cut out the plastic rectangle. This approach is more time consuming, however.

Your display hole should be slightly smaller than actually needed. This allows for aesthetic shaping to the final dimensions, so the end product has a far more pleasing appearance.

As for the rear panel, all you need is two round holes for the power cord and fuse holder, and a series of small holes toward the top for ventilation. These can all be done with drill bits. Since it is the back side of the unit, I didn't bother to paint this panel.

Once completed, spray paint the front panel with a light color (I used yellow on the prototype). This allows the black dry-transfer lettering (which can be applied as soon as the paint is completely dry) to show up properly. After the labelling is done, spray on a light coat of clear enamel to protect the delicate lettering. Don't make this coat too heavy, though, as some of the dry transfer material will melt from exposure to excessive solvent.

Again, let the last coat of paint completely cure. Now the panel is ready for its components. Everything but the banana jacks, power switch and rotary switch should already be connected to various hookup leads, so be careful not to break these connections during handling.

However, before installing the panel stuff, let's secure the transformer and boards. I used small bolts for both of the PCBs, and longer bolts for the transformer. Due to size and space considerations, the case standoffs were not compatible, so I drilled holes in the lower half to accommodate the mounting hardware.

With everything ready, I spray-painted the case a medium blue. This, however, is a matter of personal preference. I like the yellow/blue combination, but you may wish to use other colors.

Once the transformer and boards are in place, it is now time to install the front panel elements. I started with the switches, potentiometers and banana jacks, followed by the LEDs and display.

The main board output leads are hooked to the jacks, as are leads to rotary switch S2. The meter inputs are then soldered to the proper S2 poles; one for positive and one for negative (pins 31 and 30, respectively). Also, power for the 7107 is taken from the +/- 5 volt banana jack connections. In this fashion, whenever the supply is powered up, the digital meter also receives power. It will then monitor whatever output you have selected with switch S2.

Install the back panel and wire one side of the transformer's primary to one side of the fuse holder. The other side of the fuse goes to one side of the power cord. Now run an appropriate length of ZIP cord from switch

S1 (on the front panel) to the remaining leads of the primary and power cord.

Put on the knobs, slide the front panel in place, and the system is ready to calibrate. More good news! All you need to calibrate is the voltmeter, and we will cover that next.

Calibration and Testing

The process of calibrating the DVM can be handled in one of two ways. Since we are working with a 2-volt basic meter, the reference voltage should be 1 volt, which will be set with on-board potentiometer R12. This first approach involves monitoring the voltage at pins

Photo 2-2. The completed power supply in its case.

35 and 36 while adjusting R12. When 1 volt is reached, the meter is properly calibrated. However, this method can be a bit awkward, so let's look at a second (and perhaps easier) option.

Here, we set switch S2 to one of the fixed voltages, say +5 volts, and take an independent reading at the +5 volt banana jack. Do this with a multimeter. Now, adjust R12 until the display reads as closely to the independent reading as possible. In reality, either method will do just fine for our purposes. When the second approach brings the display to the proper reading, you will find the reference voltage very near 1 volt.

At this point, the display should be reading the expected voltage, and the output indicator LEDs should be lit. Try running switch S2 through its range to verify each power supply stage. If the adjustable supply LEDs are not illuminated, advance the potentiometers. Once the output hits the LED threshold, the diode will light.

If all checks out, your supply is ready to use (see *Photo 2-2*). Close the case and envision all the neat stuff you can do with this marvelous piece of electronic wizardry. To help you out, let's take a quick look at some applications.

Applications

The power supply (*Photo 2-3*) can be of service both on and off the test bench. In addition to providing needed power for your projects and prototypes, it has a number of other useful purposes. Just to name a few: battery charging, battery substitution or checking electronic equipment intended for automobile use, such as that new tape deck, digital radio or portable cellular phone.

In essence, anything that requires a stable DC voltage within the capability of our supply can be run off our supply. This, of course, is where the variable outputs come in handy. If you need 9 volts, you can crank in 9 volts.

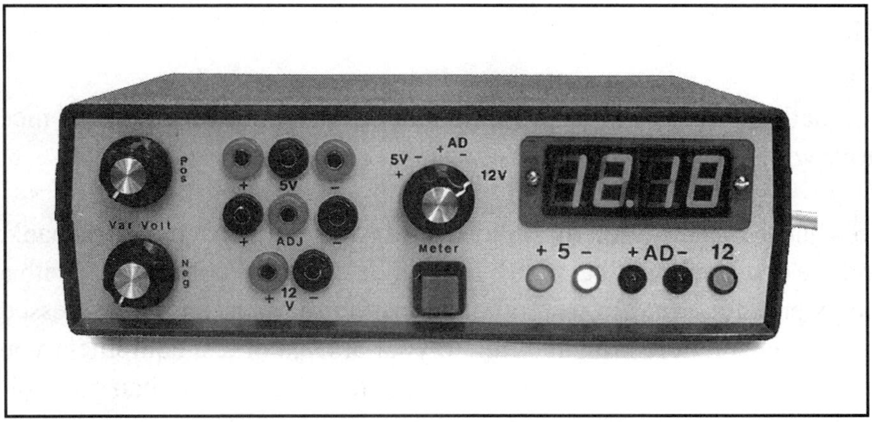

Photo 2-3. The completed multi-output power supply.

As for the test bench, this system will power a vast majority of the projects you encounter or conjure up. The split 5-volt section is ideal for logic and computer circuits, while the 12-volt leg accommodates most linear or analog chips.

The adjustable supplies? They are well worth their weight in silicon. You will grow to love these beauties, or at least to like 'em a lot. Along these lines, many operational amplifiers (op-amps) and analog-to-digital (A/D) converters require split supplies. If the requisite is something other than 5 volts, the variable supplies can handle it.

Let's not forget the exotic, obsolete chips you can dig out of surplus equipment. They are fun to play with, but some of them (such as EPROMs) require a variety of voltages (+5, -12, +25, etc.). This scenario could otherwise be difficult, but your trusty bench supply will cover you.

So, the prospective application of this unit is boundless. At the risk of repeating myself, I'll again state what I said at the outset: a dependable DC power supply is a MUST-HAVE for any work/test bench. Personally, I've found very few electronic devices that function well without power.

Conclusion

So, that's the saga of the multi-output bench-style power supply. A nice story with a happy ending. Just what we are all looking for, right?

So what are you waiting for? Build this beauty. It won't break the bank, and the end result will be of vast assistance to you. Another story with a happy ending. See, I'm just full of them. Seriously, as has been stressed in this chapter, here is an addition to your arsenal of test equipment you really shouldn't try to live without; if for no other reason than good ol' convenience when you're experimenting.

Besides, think of the joy and pride you will experience when you plug your new prototype into this unit, and it fires up; "It's alive, it's alive!"

Chapter 3
A Benchtop Signal Generator & Tracer

Introduction

If you enjoy working with audio or radio equipment, this project will certainly get your attention. A signal generator/tracer can solve a multitude of problems frequently encountered when building or testing such gear. By providing a signal that can be tracked throughout a circuit, the generator/tracer allows you to identify various gremlins and purge them from a treasured piece of equipment. This applies to the audio sections as well as the IF (intermediate frequency) and RF (radio frequency) segments of the design.

Additionally, the signal generator/tracer can make your task a lot less painful. Both audio and radio circuits are notorious headaches when it comes to diagnosing trouble, unless you have this tool in hand.

With this device, audio frequency and RF signals are produced. The tracer can handle either, thus affording the unit a high degree of versatility. There are also some other nifty tricks you can do with the tracer, which I will talk about later.

So much for high-pressure salesmanship. Let's get to the good part: understanding and building the signal generator/tracer.

Theory

Starting with the very basic theory of this device, you can introduce a known signal at any given place in the design under test, and see where

SIGNAL GENERATOR/TRACER PARTS LIST

SEMICONDUCTORS

U1	LM386N-3 Low Voltage Audio Amp, 500mw/9V
U2	CD4069 Hex Inverter
U3	7812T 12 Volt Positive Regulator
D1, 2	1N34 Germanium Diodes (Radio Shack)
D3, 4	1N914 Silicon Signal Diodes
LED1	Green Jumbo T-1 3/4 Light Emitting Diode
XTAL1	3.579545 Megahertz Quartz Crystal

RESISTORS

R1	10,000 Ohm Panel Potentiometer
R2	10 Ohm 1/4 Watt Resistor
R3	22,000 Ohm 1/4 Watt Resistor
R4	1,000,000 Ohm 1/4 Watt Resistor
R5	1,000,000 Ohm Panel Potentiometer
R6	10,000,000 Ohm 1/4 Watt Resistor
R7, 8	1,000 Ohm 1/4 Watt Resistors

CAPACITORS

C1, 2	0.001 Microfarad Ceramic Disk Capacitors
C3	0.05 Microfarad Ceramic Disk Capacitor
C4, 8	220 Microfarad Electrolytic Capacitors
C5	0.01 Microfarad Mylar Capacitor
C6, 7	7 Picofarad Ceramic Disk Capacitors
C9	1 Microfarad Electrolytic Capacitor

OTHER COMPONENTS

J1, 2, 3	RCA Panel Jacks
J4	1/8 Inch Phone Jack
S1, 2	SPDT Slide Switches
S3	SPST Slide Switch
S4	SPST Rotary Switch on Potentiometer R1
M1	0-1 Milliamp Panel Analog Meter
SPK	1 1/2 Inch 8 Ohm Speaker

MISCELLANEOUS

PCB Materials, Case, Solder, Hook-Up Wire, Hardware, Paint, IC Sockets, Knobs, Dry Transfer Lettering, Wall Plug Power Supply, Bus Wire, Sheet Plastic, Etc.

it goes and what happens to it. In this fashion, usually very quickly, the circuit's performance (or lack of it) becomes apparent.

With amplifiers, the introduction can be at the input, the output of the preamplifier, or any stage of the power amplifier. In turn, the tracer will give you either an audio response or an analog meter reading that indicates the circuit's condition.

With radios, televisions or other communications gear, signal injection can be at the antenna, detector, IF stage or amplifier. Again, the tracer's speaker or meter will tell the tale. Other devices, such as telephones and intercoms, can also be tested using this piece of equipment and the appropriate technique.

These are just some of the areas in which the generator/tracer will be beneficial. More will be said about that in the operational section of this chapter.

So, with the primary functions of the unit understood, let's take a look at how to go about putting it together. In actuality, this device is comprised of three separate circuits that share a common power supply, each of which serves a special purpose.

Starting with the *generator* side, we have two oscillator/clock configurations. Both are constructed from inverter gates, and both produce square waves. The first has an adjustable frequency output in the audio range, while the second produces a single, crystal-controlled signal in the RF spectrum.

For the adjustable version, as seen in *Figure 3-1*, three gates of a complementary metal oxide semiconductor (CMOS) CD4069 hex inverter are configured as an oscillator. Potentiometer R5 and capacitor C5 control the frequency, while oscillation feedback is accomplished by sending output pin 4 back to input pin 1, through C5 and resistor R4.

The third gate, U2C, merely inverts the clock's output to a *positive-going* signal. RCA jack J2 accesses the signal, and switch S3 allows its application to the speaker. This latter action lets you to hear the pitch while engaged in audible tracing.

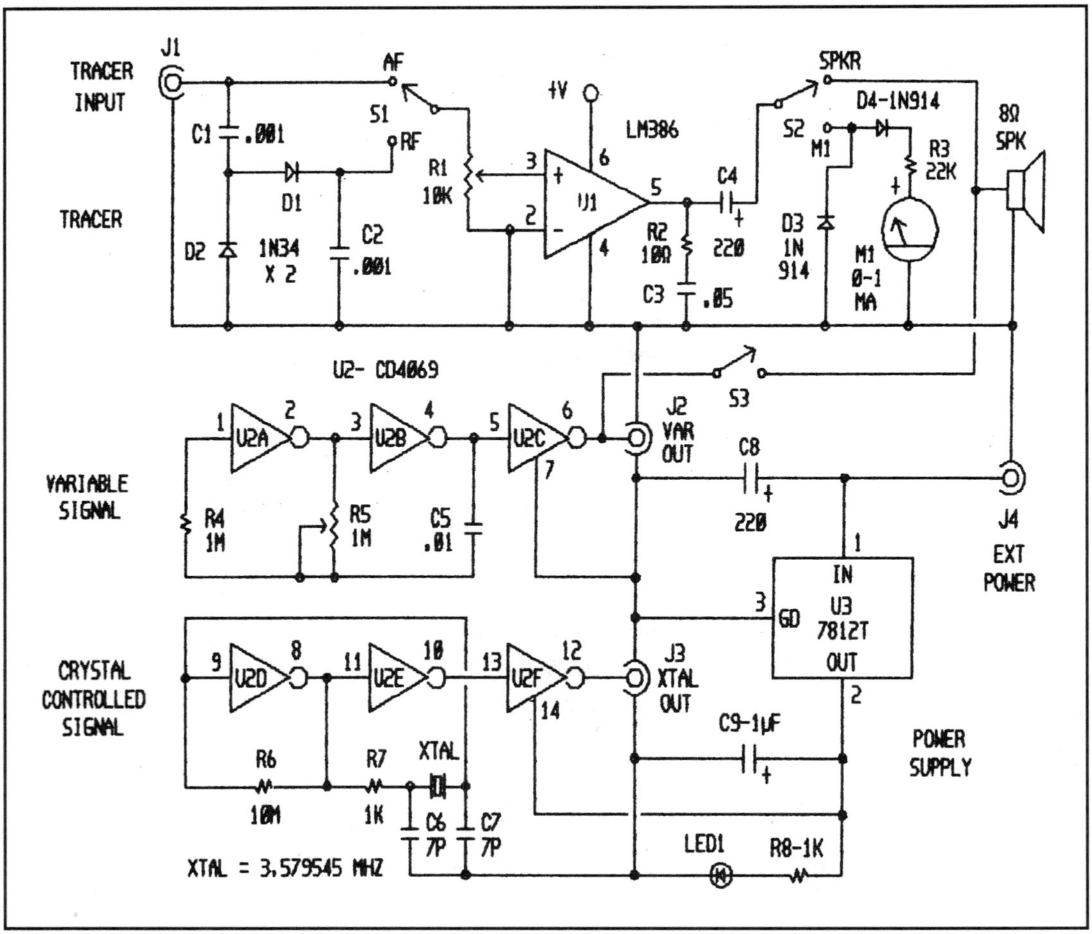

Figure 3-1. Signal generator/tracer.

The frequency can be calculated with the formula Freq = 1.4RC, where R is the potentiometer and C is capacitor C5. However, with the recommended values, I found the range to be from about 40 hertz to 4 kilohertz. This furnishes signals from a very low groan to a moderately high squeal. Hey, you can't ask for more than that, can you? Anyway, this section will provide a stable and adjustable audio tone that you can trace throughout whatever it is you want to trace.

The next section will provide a radio frequency source. Again, three gates of a CD4069 hex inverter are employed to accomplish this task, and they just happen to be the remaining three U2 gates. Here, U2D and U2E set up the oscillation, and U2F inverts the signal. Feedback comes from the crystal XTAL1/capacitor C7 junction returning to U2D's pin 9 input, while resistors R6 and R7 provide circuit biasing.

The crystal sets up the frequency at which the clock will operate, and capacitors C6 and C7 are used to jump-start XTAL1. Without the capacitors, these oscillator designs can sometimes be a little stubborn about starting.

RCA jack J3 is the signal access, and since the clock is functioning at an RF level, no provision is present to direct the output to the speaker. After all, you can't *hear* 3.579545 megahertz. Even if you could, the speaker can't reproduce it.

You may be asking, "Why did he use all square waves?" Actually, there are several reasons. First, SIMPLICITY! You know, KISS (Keep It Simple, Stupid!). By using inverter gates, both circuits are drastically less complex than many other oscillator styles, especially when you start delving into sine wave generators.

Second, these clocks are highly stable, even when using a resistor/capacitor (R/C) tank. While a slight drift in frequency is acceptable, any substantial change can foul up your results. Trust me, you don't want that. As for the crystal-controlled section, this will be right on target. The drift here will be in the 1- or 2-hertz range at the worst. Incidentally, this oscillator will function with crystals from near DC (if you can find one that low) to as high as 6 or 7 megahertz. I chose the 3.579545-megahertz value because it is the standard NTSC (National Television Standards Committee) frequency, and it comes in handy when working with video equipment.

The last reason for using square waves is practicality. I can think of very few instances in which you will absolutely have to have a sine wave. In fact, the few occasions that warrant one would involve laboratory-level testing, usually well out of the realm of us hobbyists. However, a possible modification to this design could be a sine wave generator. I will leave this up to you.

The next section of our handy-dandy, nifty-thrifty marvel involves the tracing circuit. Basically, this is an amplifier with some input conditioning (for RF) and an analog meter added to the output. Nothing fancy, but it does the job.

Again, referring to *Figure 3-1*, notice that I built this segment around an old standby: the LM386 audio amplifier chip. Input is provided by RCA jack J1, and is routed by switch S1 either directly to the 386 or to a detector for RF signals.

The detector is comprised of diodes D1 and D2, and capacitors C1 and C2. C1 acts as an input coupler and DC block, while C2 handles bypass duties. D1 and D2 provide full-wave RF detection, and are of the *germanium* variety. This is important, as standard silicon diodes will not work here.

Potentiometer R1 furnishes volume control by limiting the input to U1. R1's *wiper* is connected to the noninverting input (pin 3) of the LM386, and the inverting input (pin 2) is tied to ground.

Output comes from pin 5 and is fed to coupling capacitor C4. Also connected to the output is the resistor R2/capacitor C3 series network. Often referred to as a *snubber network*, this arrangement offsets the speaker's inductance and keeps the audio output clean and stable.

Switch S2 directs the output to either the speaker or the analog meter. For the meter, diodes D3 and D4 are wired as a full-wave bridge to rectify the amplifier's alternating current (AC) output. These diodes can be the good ol' silicon types.

Resistor R3 acts as a *dropping* resistor to control the power to the meter. This prevents the dreaded *pinning* of the delicate pointer. Don't laugh. I once saw an *overload* potential literally wrap the meter's pointer halfway around one of the stop posts. Not a pretty sight!

The last section involves the power source. Here I selected my old favorite; the linear supply. Actually, this is one occasion in which I could have gotten away with not using a regulated power supply; but since the 7812T (U3) is a fifty-cent item, I decided to include it anyway. What could it hurt?

The generator/tracer can also be operated off a battery source, but the LM386 (being a power amplifier) does draw some current. So, unless you need portability, I suggest using the AC-based system.

The external voltage is applied through jack J4, and can come from any *wall plug*-style DC battery eliminator rated for 14 volts or above. You can probably get away with one marked 12 volts, since most of these units output a voltage somewhat higher than their rating.

Capacitor C8 does the input filtering honors, while C9 provides output bypass and transient filtering, and prevents U3 from oscillating. It sounds ridiculous, I know, but linear regulators often do just that.

The combination of resistor R8 and LED1 furnish an ON indicator. This is nice, but not essential, especially if you're planning on battery operation.

There it is, at least in theory: the signal generator/tracer. Our next step is to put it together, so let's have at it.

Construction

Figure 3-2, the printed circuit board (PCB) foil or solder side pattern, is provided for your convenience. The circuit can be constructed on perf board, but I prefer to use the more reliable PCB method. *Figure 3-3* shows the parts placement. While exact for the PCB, it can be used as a rough guide if you go with point-to-point wiring.

Start by installing the three jumpers and the two IC sockets. Next come the discrete components, the regulator IC and the diodes. It is best to leave the semiconductors for last, as they are heat sensitive. With that done, the board is stuffed (see *Photo 3-1*).

Now, connect appropriate lengths (lengths that will easily reach the panel components) of hookup wire to the various PCB pads. *Figure 3-3* iden-

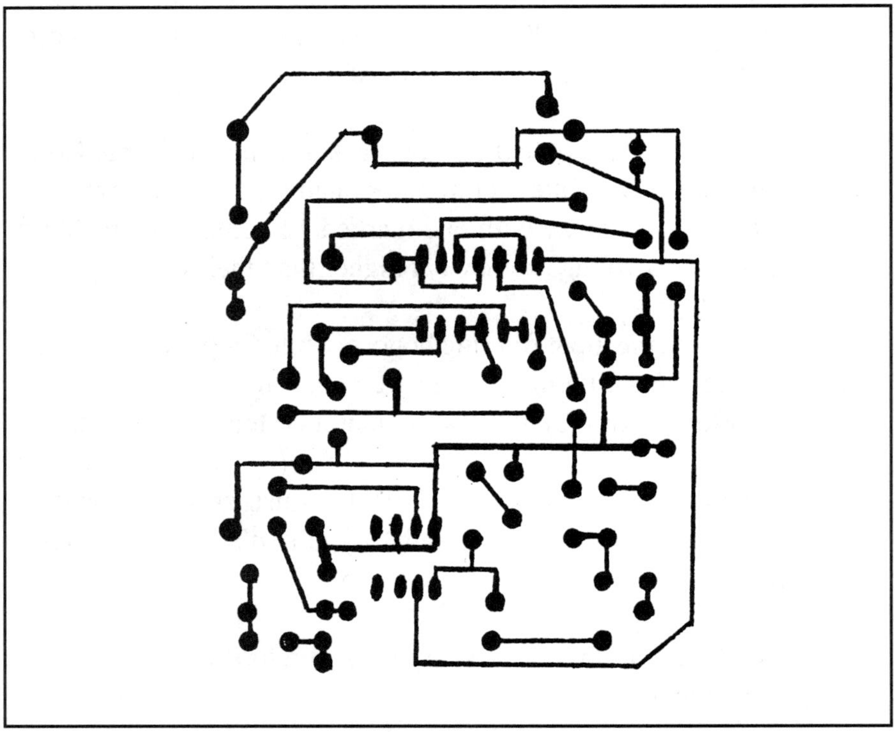

Figure 3-2. Signal injector/tracer foil pattern.

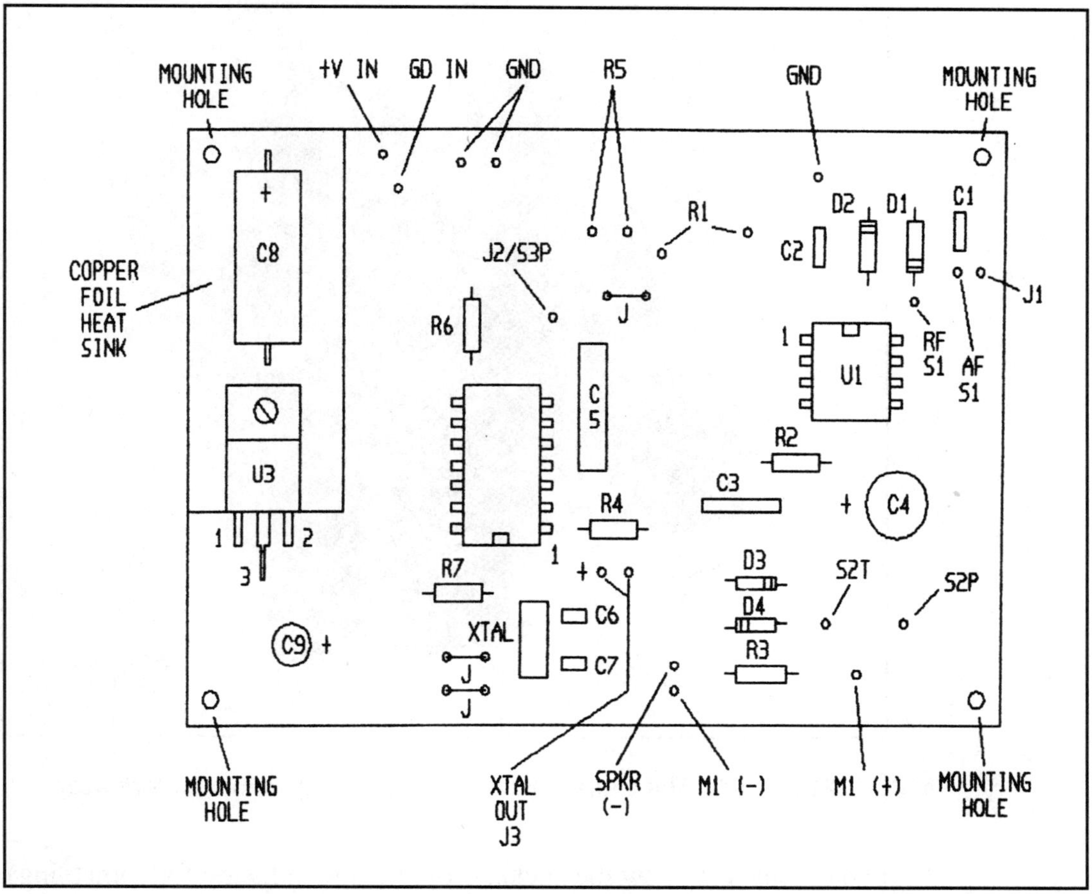

Figure 3-3. Generator/tracer parts placement.

tifies which is which. At this point, temporarily connect the external power or battery and check U3's output (pin 2) for positive 12 volts. Next, test U1's socket at pin 6 and U2's socket at pin 14 for the 12 volts. If all is well, disconnect the power source.

You can now solder the hookup leads to potentiometers R1 and R5, the meter, switches S1, S2 and S3, the speaker, jack J4 and LED1. Hold off on jacks J1, J2 and J3 for now, as they have to be mounted on the front panel before soldering.

Set all of this aside for the moment; it's time to address the enclosure. I used the wooden bottom of a case I had lying around for at least

Photo 3-1. The stuffed circuit board of the signal generator/tracer.

twenty years (I knew the sucker would come in handy for something), and I cut a plastic piece for the panel. The case began life as a toy electronic organ, with dimensions of 14-3/4" (376 mm.) by 5-3/4" (147 mm.) by 1" (25 mm.). Needless to say, the base was a little too long. So, I cut off one of the rounded ends, removed a six inch section and glued the end back on. This created an 8-3/4" (223 mm.) by 5-3/4" (147 mm.) by 1" (25 mm.) case that works nicely. The only hole required in this section of the case is for the external power supply jack. With my prototype case, due to the thickness of the wooden sides, a 3/4" hole was needed, large enough for the entire jack to fit inside.

The jack itself should be secured to a 1/16" thick plastic plate that is fastened to the back of the case with small screws. Once I did this on the prototype I gave the base a coat of blue paint and set it to one side to dry.

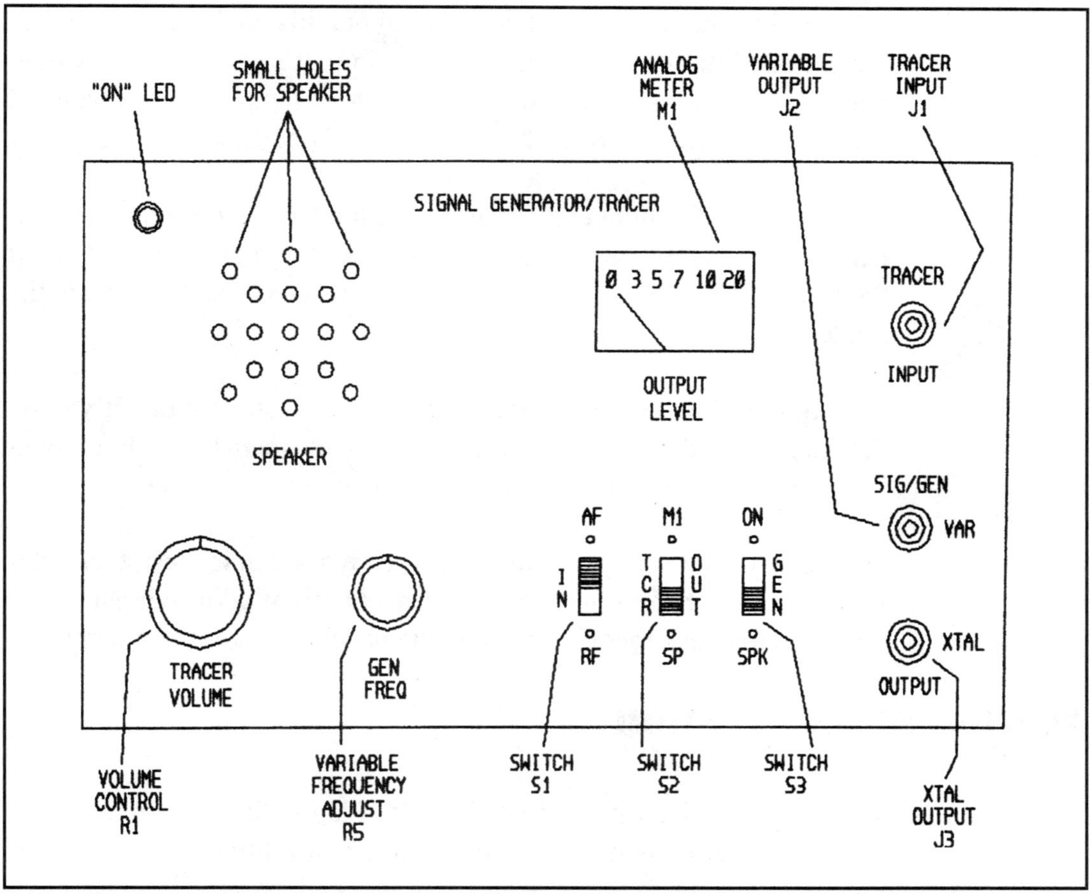

Figure 3-4. Generator tracer template.

Cut the top of your enclosure or panel from a sheet of 1/16" plexiglas. First, drill/cut all of the necessary holes for the panel components, then give it a coat of paint (I used yellow). Watch this plexiglas stuff when drilling as it can be quite brittle and will crack easily.

Next, label the panel with dry-transfer lettering. *Figure 3-4* can be used as a guide for this process. It replicates the labelling of the prototype. Apply a clear coat of paint to protect the lettering, but keep it light. Dry-transfer material (at least some of it) has an infuriating habit of melting when it comes in contact with paint solvent; a heavy coat can result in a mess. I know, I have warned you about this before, and I will continue to remind you throughout this book; but believe me, it's not that I think you're stupid or anything. It's just that I'm trying to help you from making a mistake I've made in the past. It's a pain in a certain body part to correct this error once it's been committed.

Anyway, with the light clear-coat dry, install the various panel components. As stated before, the RCA jacks have to go on first, soldered to the leads. Secure the meter and speaker with epoxy and the LED with a holder.

Speaking of the *ON* LED, connect dropping resistor R8 directly to the diode. If desired, it could be incorporated on the main board, but I found it just as easy to solder R8 to one lead of the indicator LED.

Once everything is in place, the unit is ready for a quick test, then off to work it goes. There is nothing inside that has to be adjusted; but in case you need to check the wiring, leave the panel loose for the moment.

Testing and Operation

Testing this device isn't all that difficult. Plug in the unit and rotate R1 to activate power switch S4. You should hear a little static that grows louder as you rotate the potentiometer clockwise. This is an excellent indication the tracer is working; however we will check it further down the line.

Next, close switch S3; you will hear an audio tone at the speaker. It is a good idea to set S2 at *M1*. This disconnects the speaker from U1's output. Adjust potentiometer R5, and the tone will change. This confirms that the variable signal generator is functioning.

To test the 3.579545 megahertz signal, a frequency counter or an oscilloscope will be needed. Never fear if you don't have one, however; you are going to be building a frequency counter in Chapter 11. It will give you a chance to test both the RF generator and the counter.

As a last test of the variable generator and the tracer, feed the tracer with the variable output. Be sure to first open switch S3, though. If both are cooking, you will hear the tone. Naturally, as you change the tone, the change will also be heard. If all is well, the generator/tracer is ready. Let's take a quick look at the panel functions, then discuss how to put our unit through its paces.

Most everything on the front panel is fairly obvious in nature, but I will go over all of it just to be sure everything is clear. The *TRACER VOLUME* is just what you would expect; a volume control for the tracer section. It also contains *power on* switch S4.

GEN FREQ handles the adjustment of the audio signal generator's output frequency. Switch S1, marked *IN*, determines whether the tracer input is set to detect audio, *AF*, or radio frequency, *RF*.

Switch S2, marked *TCR OUT*, sends the tracer output to either speaker *SP* or analog meter *M1*, while switch S3 *GEN* simply connects the variable generator output to the speaker. This is included primarily to allow you to hear the tone before you start a trace, making it easier to recognize the signal as coming from the generator and not another source.

The three RCA jacks are the IN/OUT ports of the device. The input to the tracer is the top jack, labelled *TRACER IN*, while *SIG/GEN OUTPUT* is the two generators. They are marked *VAR* for the variable clock, and *XTAL* for the crystal-controlled system. The analog meter is designated *OUTPUT LEVEL*, and is included as a visual indication of the tracer signal. Last, the LED in the upper left hand corner is LED1, or the *ON* display.

With all that said, let's try out our device. First find a circuit you want to test (for practice). If it is a properly-functioning radio or amplifier, normally the injected tone will be heard at the tested circuit's speaker. If the tone is not heard, you have a problem, and the generator/tracer will come to the rescue. Before I forget, the ground lead (which is common to all sections of the tester) must also be connected to the ground of the unit being tested.

For the sake of this discussion, let's say we are testing an AM/FM radio. Apply the variable generator's output to the antenna of the radio. When tuned to a dead spot, if you hear the tone at the receiver's speaker, the radio is probably working correctly. Note that with some radios, you will have to introduce the signal within the detector stage. This is all well and good regarding the signal injector, but it doesn't let us play with the tracer.

Open the receiver and probe various points within the circuit. If you can identify the detector, IF amp and power amplifier sections, so much the better. If not, then try to determine output points. These would include transistor collectors and positive-level pins on integrated circuits. Of course, this can be a hit-and-miss proposition, especially with ICs; but you can't hurt anything, so don't be afraid to poke around.

When you find a properly functioning output, you will hear the tone at the tracer's speaker. Depending on the strength of the output, the signal might be quite weak, but that's what the volume control is for.

Now attempt to locate the next stage of the circuit and test likely points there. Again, the tone will be detected at the *good* outputs. With IF sections, you may want to try changing to the tracer's RF setting. Diodes D1 and D2 will usually convert a radio frequency signal to an audible tone. In this fashion, you will be able to track the generator output throughout the circuit. If there is a problem, the lack of a tone will help you isolate a defective stage. Once you have this information, you have come a long way toward a viable remedy.

When using the meter instead of the speaker, the tracer can give you relative voltage levels within the circuit under test. I chose a value for resistor R3 that produces about a half full-scale meter reading when connected directly to the generators. So, in terms of gain, this can be a benchmark during testing. Also, since different stages within a circuit typically display different voltages, the meter will provide a more accurate indication of the voltages; that is, more accurate than the volume of the detected tone.

With amplifiers, distortion and lack of gain are two dead giveaways that trouble exists. Again, the introduction of a signal at the amp's input is the starting point. From there, the signal can be tracked through the various stages.

Having the schematic diagram greatly assists in testing a circuit; however, the device can be checked without it. Of course, inputs or antenna connections are easily identified, but regarding the internal circuit, inputs and outputs are not always easy to spot.

For transistors, the base lead is usually the input point, while the collector generally handles output duties. This is because the current flows from the emitter to the collector, and is controlled by the base potential.

For ICs, the story is a little different. The best approach is to obtain the IC pinout that reveals the function of each pin. Since there are so many different layouts for so many different ICs, it is hard to provide any rule-of-thumb hints. However, as previously stated, the tracer isn't going to hurt the device, so shop around.

Some common problem areas to be aware of (and to look for) are gain, distortion and semiconductor leakage. All can result in improper circuit operation. Gain can be either too high or too low. Low gain often results from bad capacitors or leaking transistors, with the latter reducing the gain to practically zero if enough leakage is present. Emitter bypass capacitors can have the same effect. Gain that is overdriven or too high is frequently the consequence of transistor leakage. So, as can be seen, a leaking transistor can produce either type of gain difficulty.

Distortion customarily goes hand-in-hand with gain problems, although it can also be the product of incorrect biasing. Once again, transistor leakage could be the culprit. Base-to-collector leakage is the most common leakage problem and causes low gain.

In a *feedback* amplifier design, low gain will almost always produce distortion, so this should be considered when distortion is encountered. Capacitor leakage is another area to check, as it also can result in low gain that produces distortion.

Anyway, these are some of the maladies to look for when using your signal generator and tracer. As promised, there are a couple of extra duties the tracer will perform. These involve using it in the radio frequency (RF) mode. First, with the meter connected to the output and an antenna attached to the input, you have a handy field strength meter (FSM). One nice feature is that this type of circuit (often referred to as a *passive detector*) provides a wide band response. Additionally, it will not interfere with other radio equipment.

A second function is along somewhat these same lines. Connecting the speaker to the output will give you audio reception of radio signals over a wide range of frequencies. Of course, you have to be within reasonable proximity to the transmitter, but the proximity is largely determined by the transmitter's power.

Both of these tracer extras can be very useful for diagnosing amateur radio or HAM gear troubles. With much of that equipment, you won't have to be all that close to the transmitter. Hams are allowed to use up a thousand watts on some bands.

Don't forget the crystal oscillator. As stated before, 3.579545 megahertz is the standard frequency for broadcast television (at least in the United States). Sometimes known as the *color burst* frequency, all video timing is derived from it; thus, it is a natural for troubleshooting television and video circuits. The inverter clock furnishes a very stable signal. One caution, though: be careful when working inside TV sets. Some of the high voltages present will speak to you directly (perhaps painfully) if you tangle with them.

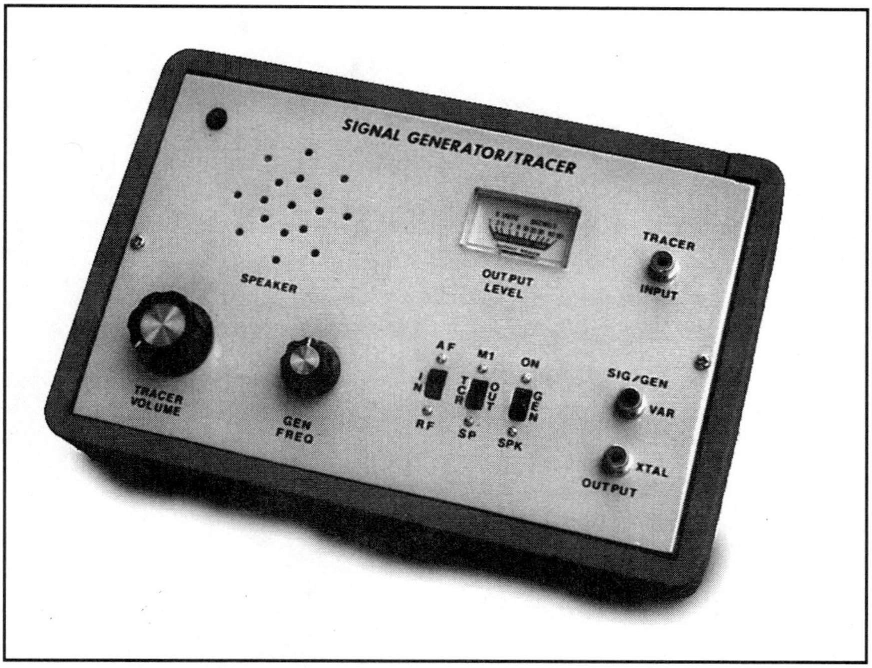

Photo 3-2. The completed signal generator/tracer, ready for use.

Conclusion

Thus ends the tale of the trusty signal tracer/generator (*Photo 3-2*). Interesting, wasn't it? Don't you love the part about how it tracks down the bad guys? Kind of like a mystery or a western. This is not the real end to the story, though. Once you have this gem at your disposal, you will be able to spin your own interesting tales. I guarantee it! The unit is easy to build, reasonably inexpensive, and with its simplicity, it will serve you for years to come. So, what are you waiting for? Build it!

Seriously, I doubt you will regret adding this device to your test bench. Like so many simple pieces of test equipment, a signal generator/tracer provides much more versatility to your experimenting and adventures in electronics. Have fun with it!

Chapter 4
A Tri-State Logic Probe

Introduction

You've worked hard. You've built a couple of tough projects. You deserve a bit of a break. So, our next endeavor is going to be a logic probe. It is somewhat simpler in nature and far easier to build than, say, the projects in chapters 1 and 2. It is also a lot cheaper! However, you will be rewarded with a piece of equipment that is very beneficial when working with digital circuits. Using two integrated circuits (ICs), this device tells you whether a test point is high (positive), low (negative/ground level) or pulsing (oscillating).

At first glance, this might not seem like an overly important project; however, you will soon learn how valuable logic information can be once you delve into the world of digital electronics. As you may already know, in the digital world, everything is either ON (high) or OFF (low), and there is no gray area in between. The advantage of this (or at least one advantage) is the simplicity of data transfer. Everything is either a *1* or a *0*, which greatly elevates both the speed and accuracy of transmissions.

Digital electronics is becoming an ever-growing influence on our lives. Of course, analog processes (a spectrum of values between 1 and 0) such as radio and TV still have their place, but even these areas are feeling a digital invasion.

In most video gear, many operations once handled by analog circuitry are now routinely managed digitally. This is both good news and bad.

LOGIC PROBE PARTS LIST

SEMICONDUCTORS

U1	74LS04 Hex Inverter
U2	74121 Monostable Multivibrator
DSP1	MAN72 7-Segment Common Anode LED Display

RESISTORS

R1	220 Ohm 1/4 Watt Resistor
R2	22,000 Ohm 1/4 Watt Resistor
R3	22 Ohm 1/4 Watt Resistor
R4	100 Ohm 1/4 Watt Resistor

CAPACITOR

C1	1 Microfarad Tantalum Capacitor

MISCELLANEOUS

Case of Your Choice, Solder, Hookup Wire, PCB Materials, Test Lead Wire, Alligator Clips, Epoxy Cement, Probe Tip, 14-Pin IC Sockets (2), Etc.

Digital is more precise and usually easier to troubleshoot, but to do so you will need some new test equipment. This is where our logic probe comes into play.

A logic probe's ability to check the levels within digital circuits helps you to quickly diagnose problems, including bad gates, counters, converters and so forth. You will have a well-equipped test bench when you finish building all the devices in this book, so the ol' logic probe just has to be there.

Theory

As stated previously, the job of a logic probe is to determine the level of any given point within a circuit. The probe can be as simple as a transistor and a couple of resistors (see *Figure 4-1*) or, in varying degrees, considerably more complex.

Probes are designed to distinguish only high signals, only low signals, both high and low signals, or the presence of a pulsing signal in addition to high and low. They can use regular light-emitting diodes (LEDs) to exhibit their results or, as in the case of this project, call a 7-segment display into service. Probes can be made from counters, timers and the like, but usually rely on inverters to perform their tasks. Pulse signals are typically detected with monostable multivibrators, yet this can be done in other ways as well.

In essence, logic probes can be developed in a substantial number of ways. I'm not going to say that any one way is better than the rest. I do have my favorites, though, and this is what we will be looking at here.

Let's get back to the fundamentals. Again referring to *Figure 4-1*, we are using transistor Q1 as a switch. As long as the signal at the probe

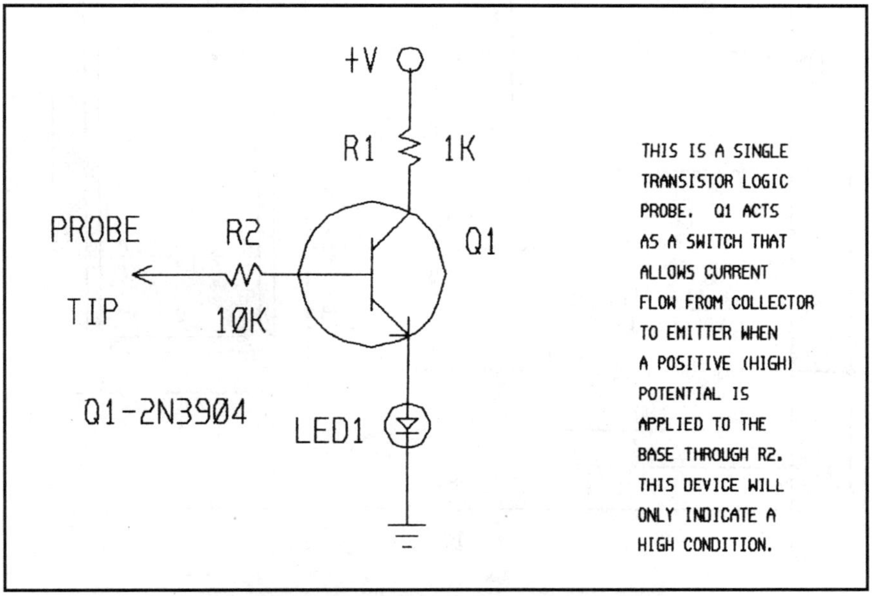

Figure 4-1. Transistor logic probe.

is low, the LED will not light. This is because the transistors's base will not allow for electrical flow from the collector to the emitter. However, when a high signal is introduced to the probe, the base junction is turned on and allows electrical flow. Thus, the LED glows. Resistor R2 handles biasing while resistor R1 acts as a dropping device to protect the diode.

This type of probe does have its unique purposes. When you get to Chapter 6, I think you will recognize this circuit as the IC pin indicators. Also, it's real handy if you want to monitor the action on computer bus lines or other logic operations. As neat as it is, though, this circuit configuration only tells us when the probe has found a high signal. This really isn't good enough for our objectives. So, let's look at a design that provides more information.

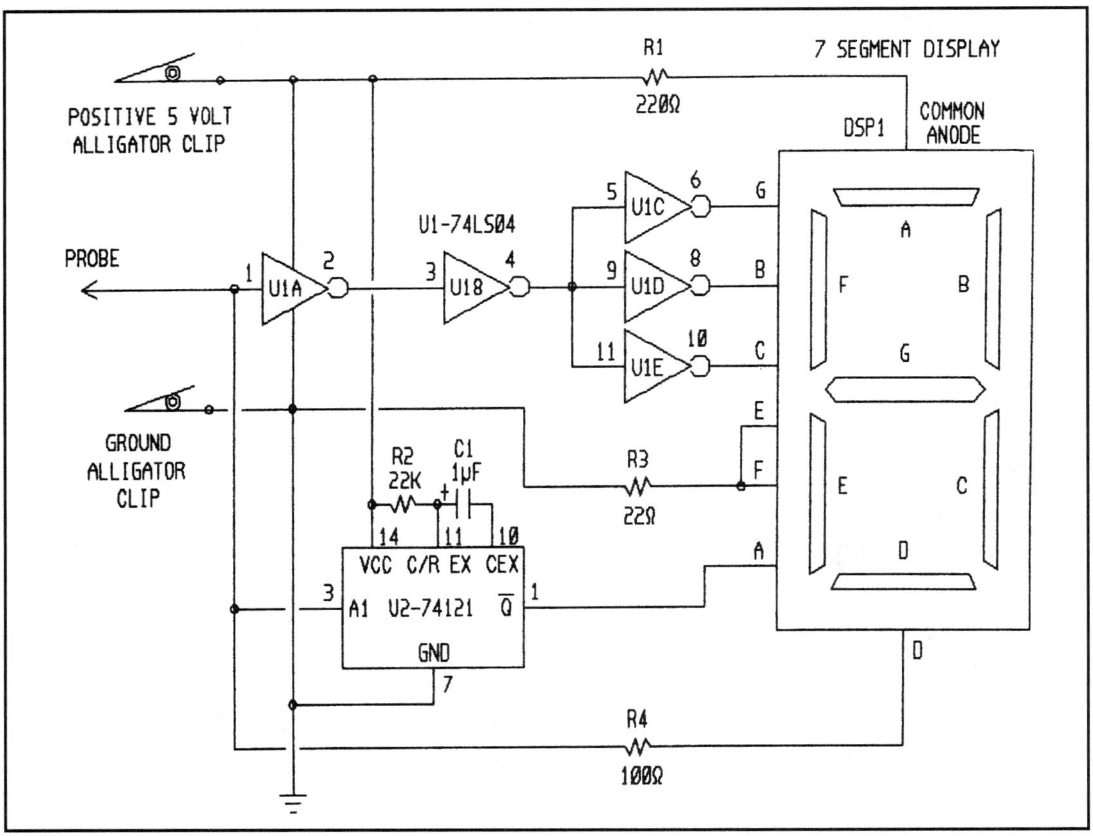

Figure 4-2. Logic probe schematic diagram.

Figure 4-2 illustrates the device we are going to build. As you can see, it incorporates two integrated circuits and a 7-segment common-anode LED display unit; along with a few discrete components, of course.

I have chosen transistor-to-transistor logic (TTL) chips, which does limit the probe to a maximum of 5.25 volts. I did this because most of the circuits you will be testing operate at 5 volts. This is pretty much a standard in the digital world.

I know, not all digital circuits use 5 volts, and you could employ the complementary metal oxide semiconductor (CMOS) IC equivalents to increase the voltage range. However, this would require a switching system to adjust resistance on the LED display. For as often as it's going to be useful, I just couldn't justify the extra complexity, work and circuitry of a switching system.

Anyway, one nice aspect of this project is that it lends itself readily to all kinds of modification. Many features of this probe depend on your own personal preferences.

Back to our project. As with all logic probes, power comes from the circuit under test. This is the purpose of the two alligator clips; to rob the 5 volts from the "sick" circuit. This not only powers the ICs, but also establishes a common ground between the probe and the device being tested.

Once the clips have been connected to the circuit's voltage, the probe or tip is then touched to the point(s) in question, and the logic level is displayed by DSP1. To understand how this is done, let's walk through the functions of the probe. Each of the inverter gates will reverse (invert) at its output whatever is sent to its input. As the signal is routed through the various gates, the level is manipulated to produce what the display segments want to see.

Since this is a common-anode display, the segments only light when a low (ground level) potential is applied. Thus, to get the display to read

H for *high*, the circuit has to produce lows on segments B, C, E, F and G. For an *L* as in *low*, segments D, E and F need to be activated.

You may have noticed that both high (H) and low (L) share the E and F segments; this helps to simplify the layout. E and F can be permanently taken low by hard-wiring them to ground through resistor R3. This keeps E and F illuminated at all times. This means you only have to worry about B, C, D and G, where the logic gates do their stuff. Actually, the D segment is controlled by the input, so we are really only concerned with the other three.

Let's say we have a low input at the probe tip. This will cause the D segment to light, and U1A will invert it to a high. U1B now takes the high back to a low and feeds it to the inputs of U1C, U1D and U1E. The latter gates take the low back to highs, preventing the B, C and G segments from glowing. So, F, E and D light up, forming the letter *L*. You with me so far? Good! Let's move on.

When the probe tip finds a high, segment D is turned off and U1A converts the high to a low. Then U1B changes the converted low back to a high, and U1C, U1D and U1E invert it to three lows. Each gate is connected to one of the B, C and G segments, and all three illuminate. This combined with the already-glowing E and F gives you the letter *H*.

Any questions? Nah, I knew you had this down. Let us now look at the probe section that acts as a pulse detector.

Since we know that the pulsing signal alternates between high and low, it stands to reason that it will turn both individual display indications on and off. The frequency of the pulse determines the speed of this change. Hence, for a very long pulse, the change between an *L* and an *H* will be slow and distinct to the eye. The presence of this alternation is, of course, one way to detect a pulsing input. However, as the pulse frequency increases, the visual distinction becomes less prominent. Eventually, you reach a point where the L and H appear to be on at the same time. This is due to the phenomenon known as *persistence of vision*.

If you're not familiar with the term *persistence of vision*, it is the ability of the brain to hold an image seen by the eye for a certain length of time before it fades away. This is what allows individual motion picture frames to materialize as smooth movement. The same principle applies to television. What we perceive as continuous motion on a TV screen is actually the product of many separate *frames* being scanned onto the picture tube. This works because the frames are presented fast enough so the brain receives the second image before the first one fades.

With a pulsing signal of about 20 hertz or above, what we see is a *U* with a hash mark (-) in the middle. Not exactly what we want. So U2, a 74121 monostable multivibrator, is employed to detect the pulse. This method provides a positive indication of an alternating input.

Monostable multivibrators are *one shot* oscillators, and are controlled by an R/C (resistor/capacitor) tank network. In the case of the 74121, pins 10 and 11 (the external capacitor and external resistor/capacitor connections) handle the R/C tank. These components determine the frequency of the 121's output pulse (length of time the output stays active) each time the chip is triggered. Triggering is accomplished on the rising (positive) edge of the pulsing signal. The incoming signal is connected from the probe tip to pin 3, the A1 input, to trigger the 74121.

The multivibrator has two separate outputs, *Q* and *not Q*. We use *not Q* (pin 1) because it is low during the 121-output cycle. Remember, our display needs low signals. Pin 1 is connected to the A display segment, and when a pulsing signal is present at the probe tip, the display reads *8 (all* segments lit). This, of course, presumes the frequency is above 20 hertz.

When you fire up the logic probe, the display will indicate a low signal, or an *L*. Why, you ask? Well, since the tip is not detecting a high, the probe assumes the signal is low or at ground level. In reality, of course, the tip is floating; but as long as you know that, all is well.

So, as a review: if you touch the tip to a point and the display does not change from *L*, you know you have a low signal. If a high is present, the display will transform to an *H*. An *8* indicates a pulse or alternating signal. It's as easy as that.

Construction and Testing

Here is where modification really gets useful. This device can easily be constructed with either a point-to-point wiring or printed circuit board (PCB) technique. I chose a PCB and have included the foil pattern (*Figure 4-3*) and a parts placement diagram (*Figure 4-4*) for the prototype I built.

However, a multitude of possible enclosures exist, and this will figure heavily regarding the circuit layout. I used a 5" (128 mm.) by 5/8" (46 mm.) section of clear red plastic pipe (an old surplus purchase) with end caps. This provides ample room for the 4-1/2" (115 mm.) by 1/2" (13 mm.) by 7/16" (11 mm.) stuffed circuit board (*Photo 4-1*). It also allows the 7-segment LED display (as well as the rest of the circuit) to be viewed clearly.

The long, narrow tube makes the design very ergonomic; but, as stated before, you may want to use a different shape or design. Old cigar tubes make good cases, as well as small remote-control boxes. The choice of enclosure is yours.

Presuming you are using the same-size case I used, I will proceed with construction per the prototype. Let me say at the outset that this is a

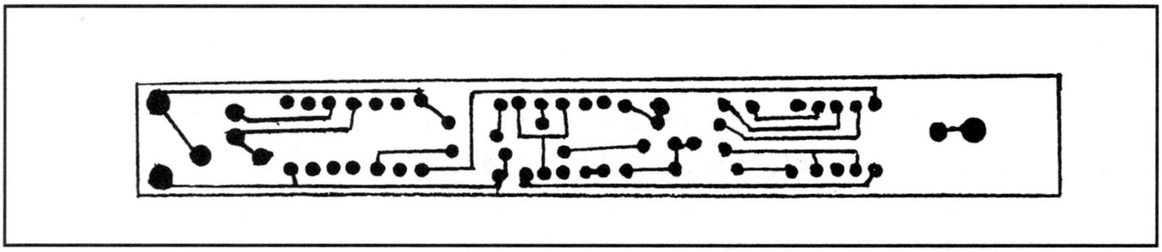

Figure 4-3. Logic probe foil pattern.

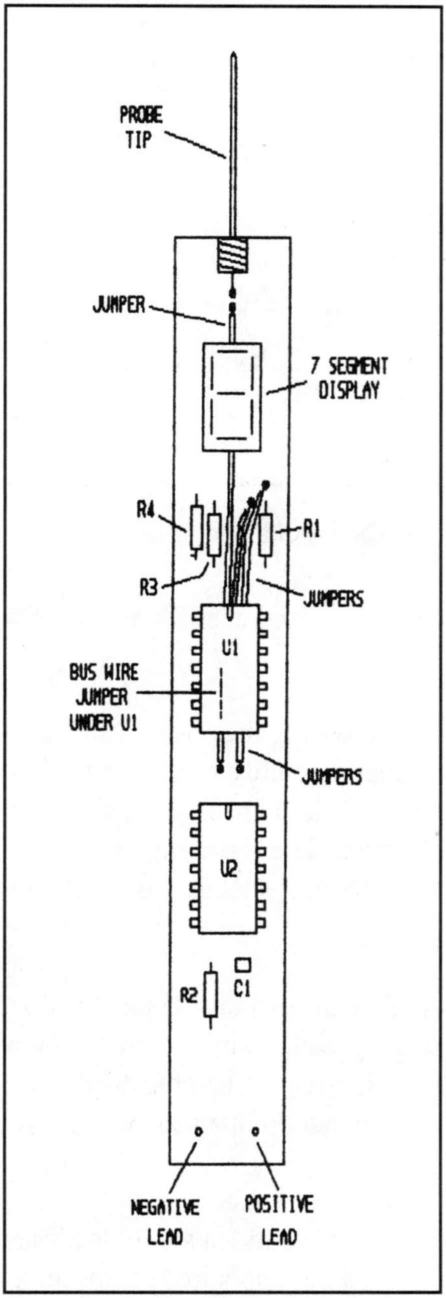

Figure 4-4. Logic probe parts placement.

tricky board. Due to its long, narrow dimensions, care needs to be taken when soldering. For the same reason a number of jumpers are required, unless you go with a double-sided printed circuit board.

The first step is to put in the six jumpers. They all run under either the display or U1, so they have to go on before the IC sockets and display. Five of the six are made from insulated wire; however, all could be eliminated with a double-sided PCB. If you are good at constructing such boards, this is as good a time as any to make use of your talent.

With the jumpers in place, install the socket for U1. The jumpers need to be bundled together and channeled down the center of the socket. They will fit, honest.

Next come the discrete components, U2's socket and the display. With that done, secure and connect the tip. For the probe, I used a needle for inflating footballs, attached

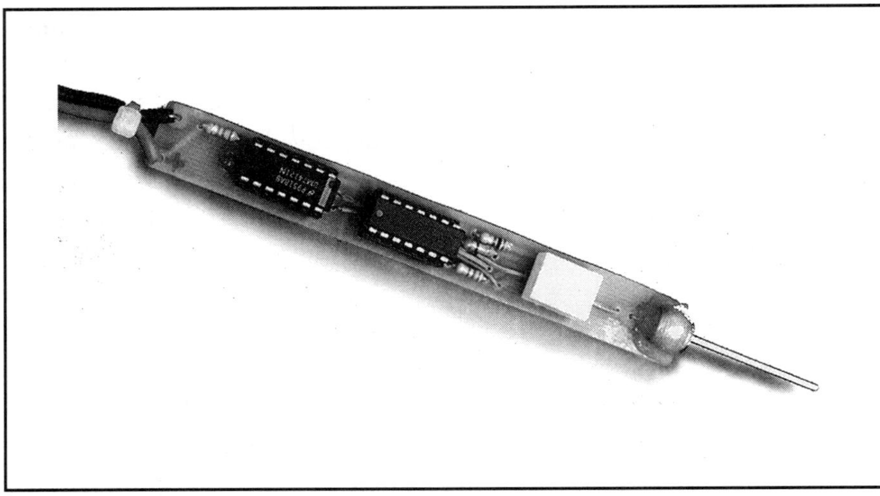

Photo 4-1. Small, thin, stuffed PC board of the logic probe.

with bus wire and epoxy cement. A nail or a piece of stiff wire will also work.

Last, temporarily solder short pieces of bus wire to the positive and negative pads. This will allow for a pre-final assembly test. For the test, hook up a 5-volt supply to the proper leads and check pin 14 on each socket against ground (pin 7) for +5 volts. If everything is all right, disconnect the power and plug the ICs into their sockets. Be careful to observe orientation.

Power up again, and you should see an *L* on the display. Touch the tip to the supply's positive post, and the display will change to an *H*. Now, rapidly connect and disconnect the tip from the post, and the display will just as rapidly change from *H* to *L*. If you can do this fast enough, you will see the *8*.

Once the test is successfully completed, remove the bus wire leads and run 24" (611 mm.) lengths of black and red hookup wire first through a hole in the back cap, then to the board's power points. Solder these, install the front cap while inserting the tip through another hole, and the probe is ready to go (*Photo 4-2*).

Photo 4-2. The completed logic probe in its clear red, plastic home.

Conclusion

Normally I would include an operations section at this point; but since we have pretty much covered how the probe operates, I'll just get us out of this chapter. I mean, using this thing ain't rocket science, but that's part of its charm.

As I said at the outset, a logic probe is a must for any work bench, especially if you plan to work with digital circuits. Fortunately, logic probes are simple to use, easy to build and cheap, which makes them just that much more attractive. Unfortunately, too often the probes fall into the "I'll build one of those when I get a chance" category, and much too often, they never get built. That's a pity as this tool will serve you well. Considering the amount of time and expense involved in the electronics hobby, the logic probe is a genuine diamond in the rough.

So, break out the parts box and your surplus stash. You can probably whip this one up in a single evening. You'll be checkin' them there logic levels in no time at all. Happy hunting!

Chapter 5
A Bipolar Transistor Tester

Introduction

Over the years, I have bought literally thousands of surplus transistors. In one of my more spectacular purchases, I paid a penny a piece for the little devils, and boy was I proud of myself. Unfortunately, they came in a large plastic bag, all tangled and twisted with no markings. It was then that I decided to build a device to both test and identify these gems. Identify, that is, NPN or PNP types.

As I thought about this project, it suddenly became clear that such a tester didn't have to be a major technological achievement. In reality, it should be fairly simple to make. After all, transistors are basically controlled-switching devices, so all I had to do was design a circuit that tested this property. Like diodes, transistors work only in one direction with (provided I didn't try to electrocute them) no substantial danger of reverse polarity damage. Hey! As long as I knew the polarity I was applying, the rest was simple.

Not only had I discovered a modest method for testing transistors, I could now take that big bag of semiconductors and find out what I had (or if I'd been had). Yes, the batch of semiconductors turned out to be somewhat of a disappointment, and a lesson; but the experience did result in one heck of a pragmatic test device.

So, here is the result of my intense research: one handy-dandy transistor analyzer. As promised, it won't strain you brain or mechanical abilities too far.

TRANSISTOR TESTER PARTS LIST	
SEMICONDUCTORS	
LED1	Green Jumbo T-1 3/4 Light Emitting Diode
LED2	Red Jumbo T-1 3/4 Light Emitting Diode
RESISTORS	
R1	470 Ohm 1/4 Watt Resistor
R2, 6	1,000 Ohm 1/4 Watt Resistors
R3	33,000 Ohm 1/4 Watt Resistor
R4	10,000 Ohm 1/4 Watt Resistor
R5	10,000 Ohm Panel Potentiometer
R7	2,200 Ohm 1/4 Watt Resistor
OTHER COMPONENTS	
M1	0-1 Milliamp Analog Meter Movement
S1	4PDT Pushbutton Switch
S2, 3	Normally Open Momentary Pushbutton Switches
BAT	9 Volt Alkaline Battery
MISCELLANEOUS	
Solder, PCB Materials, Hook-Up Wire, 4" x 6" x 1 7/8" Plastic Project Box (Radio Shack), 9 Volt Battery Connector, LED Holders, Paint, Dry Transfer Lettering, Knob, Etc.	

Theory

The *bipolar* or *junction* transistor that we've all come to know and love is comprised of three components: the collector, the base and the emitter.

At the risk of sounding simplistic, I'm going to use the following analogy to explain the device's operation. Visualize a building with people standing outside (the emitter). The base can be thought of as a door that, when opened, allows the people to move inside (the collector). Naturally, the wider the door is opened, the more people it will let in; but it takes energy to pull the door open, like someone pulling on it. I know, this reads like a grammar school math quiz. Anyway, the same principle pertains to the transistor. Without energy applied to the base, the door stays shut and the current (the people) cannot move from the emitter to the collector. However, as a potential is applied to the base, the door opens. The more potential, the wider the door opens, thus a greater flow of current. In this fashion, transistors act as switches.

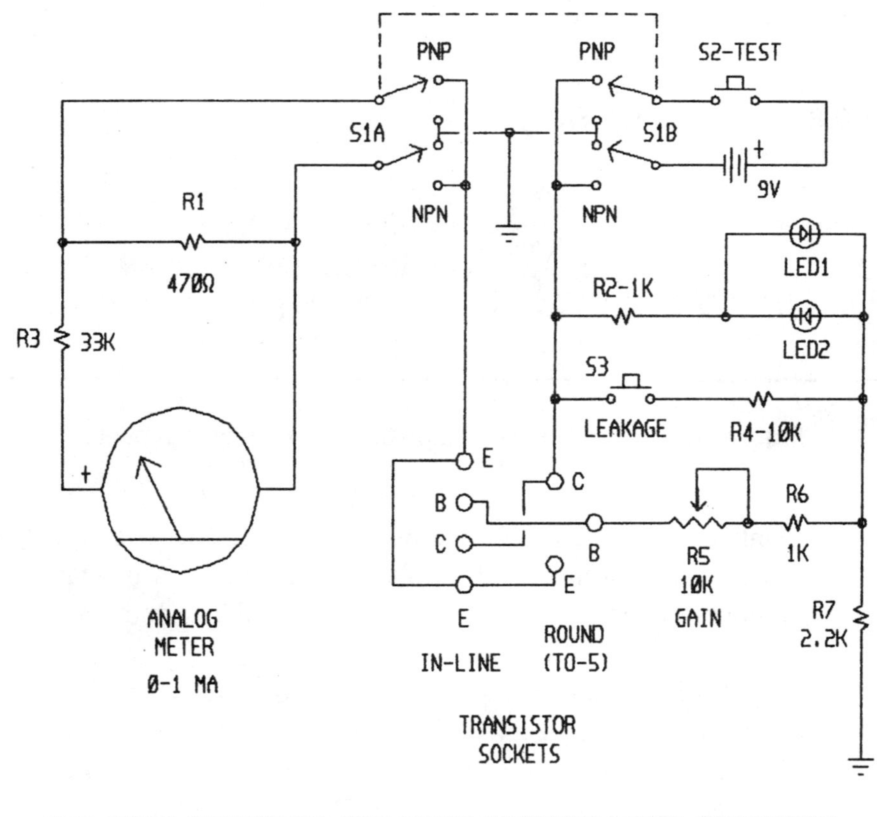

Figure 5-1. Transistor tester schematic.

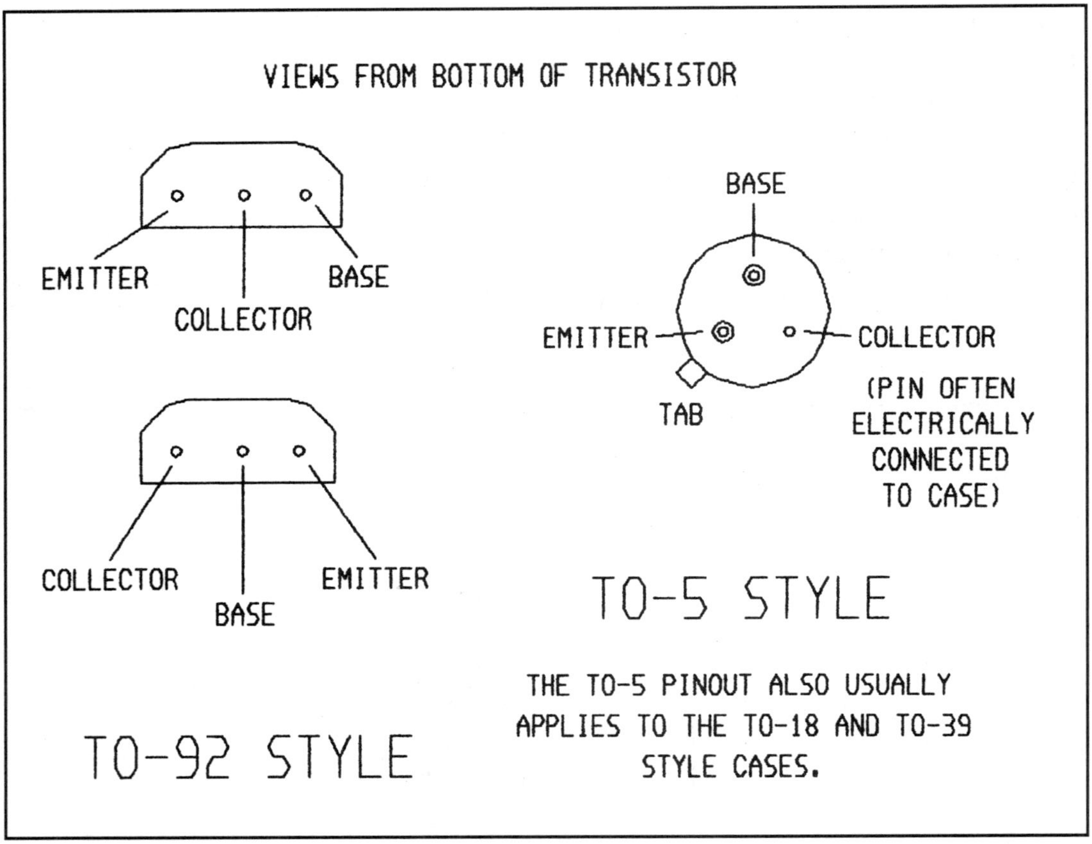

Figure 5-2. Some of the common transistor pin assignments.

Unlike a standard mechanical switch, the transistor can control the amount of flow by varying the base voltage. Again, this is analogous to the amount of energy expended to open the building door.

So, this is also how transistors amplify; by using a small potential at the base, they can move a much larger potential from the emitter to the collector. In essence, you have *amplified* the small base potential.

Perusing *Figure 5-1*, you will notice that switch S1 selects the type of transistor to be tested (NPN or PNP). It does this by sending the proper polarity to the collector. For example, when checking a PNP unit, the collector receives the positive side of the battery. For an NPN device, the polarity is reversed, sending the negative battery connection to the collector.

The base potential is derived via resistors R6, R7 and potentiometer R5. R5 acts as a gain control that lets you vary the potential. This action gives a meter indication of the emitter-to-collector transfer, demonstrating proper operation.

Speaking of the meter, it is connected to the second half of switch S1. Again, this is for polarity control. As seen in *Figure 5-1*, the emitter of the transistor under test is routed to an analog meter, through voltage divider R1/R3.

Since the emitter polarity changes depending on the transistor type, without switch S1A, the meter would read in reverse when testing NPN devices. This is not only bad on the meter, it won't give us the information we are trying to obtain. Thus, the trusty switch keeps the meter reading correctly.

LED1 and LED2 are included as visual indicators of the type of transistor the unit is set to test. They are wired from two of the S1B poles,

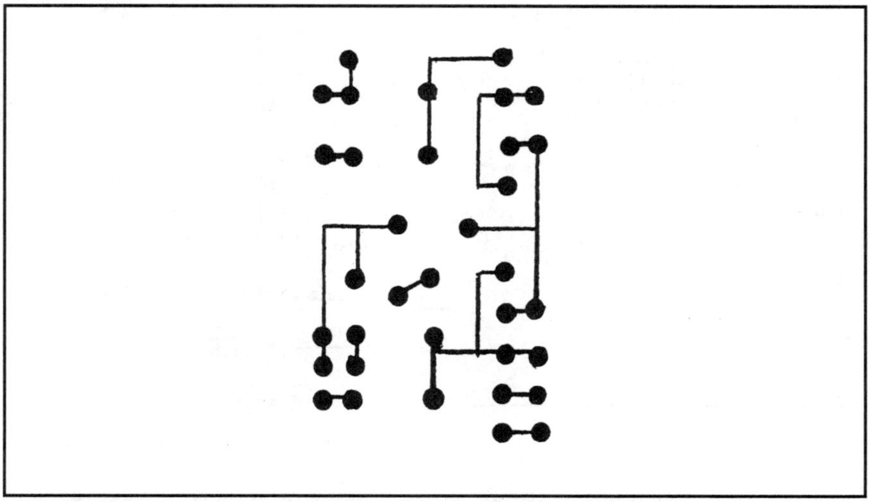

Figure 5-3. Transistor tester foil pattern.

through dropping resistor R7 to ground. The diodes are also wired opposite each other, allowing LED1 to light when *PNP* is selected and LED2 to glow for *NPN*.

Switch S3 provides a leakage test by dropping the collector potential to ground through resistor R4. Here, a small rise in voltage is normal while a large rise indicates a bad transistor, or at least one that is heavy in leakage. Switch S2 is the test switch, and merely applies power to the circuit.

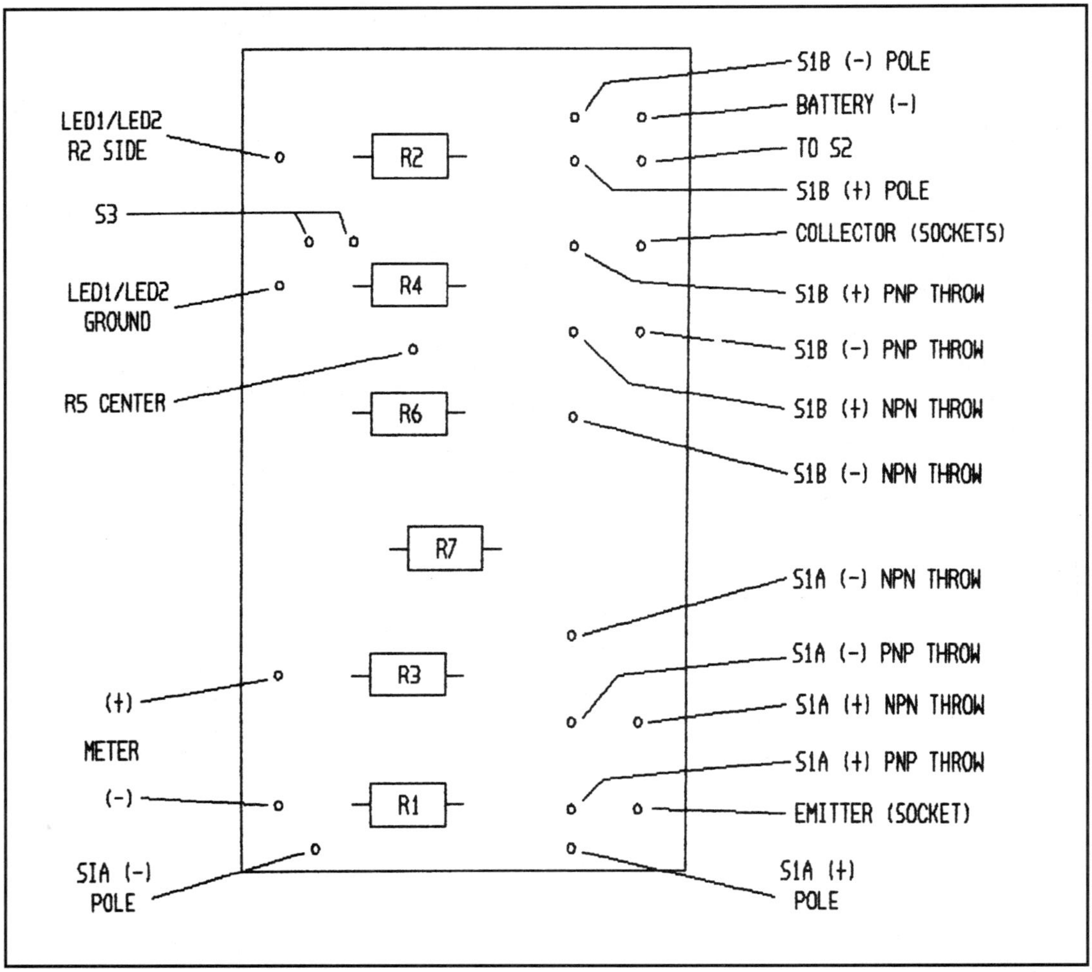

Figure 5-4. Tester parts placement.

The last feature to consider is the transistor sockets. I incorporated the two most commonly employed configurations. *Figure 5-2* illustrates the pinouts for both the *in-line-* and *round*-style devices. You'll note that the in-lines come in two different flavors; one with its base in the center, the other with it on the outside. Between these two sockets, which are wired in parallel, you should be able to test the majority of contemporary transistors. However, some of the older or more wacky arrangements might take a little lead juggling. You know the saying, "Nothing's perfect."

So, that's it. Pretty simple, eh? I tried to make this project complicated, but that's a hazard when striving to overkill electronic projects. You know, "Gee, it's electronics. It's got to be tough." Anyway, now you know how the tester works, so let's put it together.

Construction

For this one, the circuit board (such as it is) can be either point-to-point or PCB. *Figure 5-3* illustrates the foil pattern for a printed circuit board, and *Figure 5-4* shows the parts placement. However, a piece of perf board and some hookup wire will easily do the trick.

Since most of the components are mounted on the front panel, the board is the easy part. Install the resistors, and set the board aside for the moment.

For the case, I used a plastic project box 3" (76 mm.) by 6" (127 mm.) by 1-7/8" (48 mm.) in size. This enclosure accommodates everything, with room to spare. Actually, it is larger than needed, but this makes the tester easier to handle.

Using the template in *Figure 5-5*, cut and drill the necessary holes for the panel components. Next, spray paint the front panel (case top) with a light-colored enamel. I always use yellow, but any color will do. The primary concern here is keeping the black dry-transfer lettering legible.

On that subject, once the paint has dried it is time to label the panel. Again, *Figure 5-5* can be used as a guide. The last step is a clear coat to keep the lettering from being damaged. Dry transfer is great, but it is also delicate.

With the clear coat dry, install the switches, meter, LEDs, sockets and potentiometer. Regarding the sockets, I used single in-line pins (SIPs) arranged in the shapes I wanted, but commercial sockets are available and could be employed. Last, hook suitable lengths of lead wire to the board, and you're ready for the test procedure.

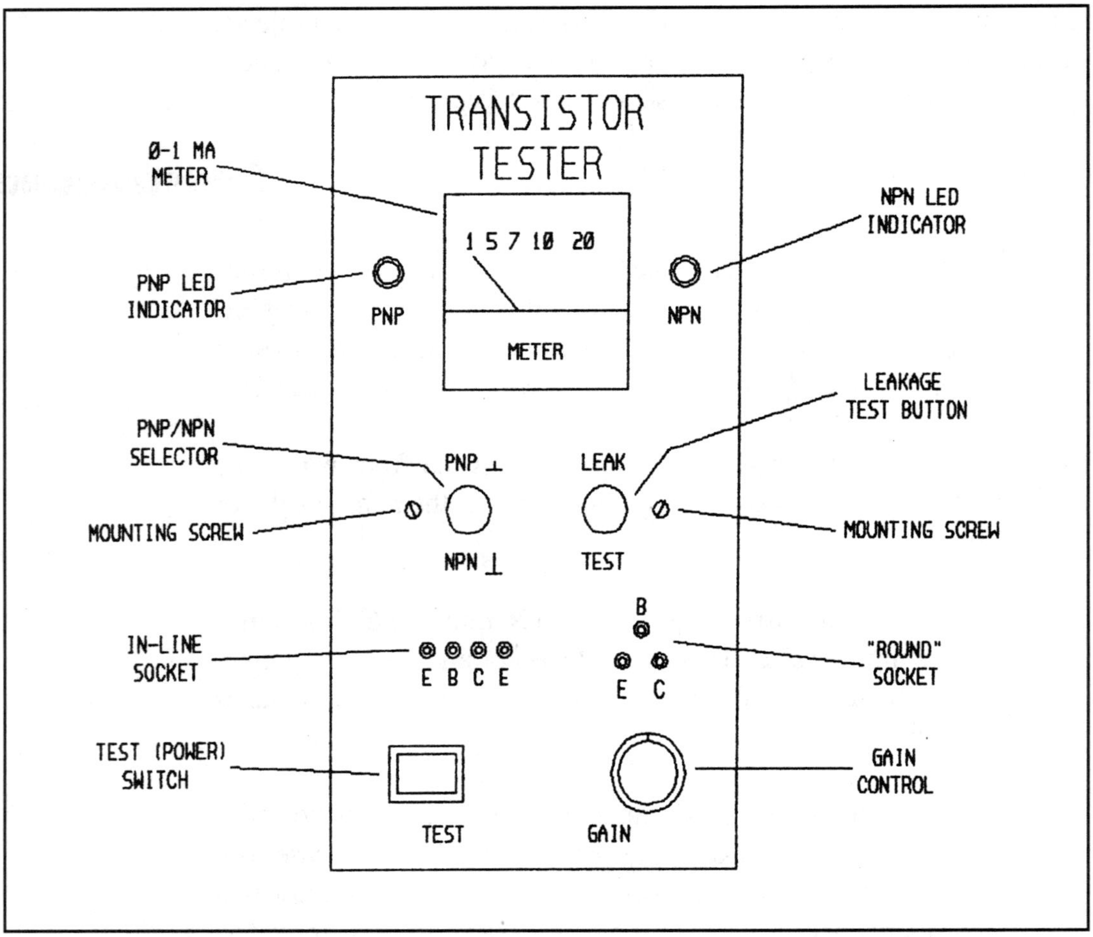

Figure 5-5. Transistor tester template.

Testing and Operation

Testing the tester is about as simple as designing it (*Photo 5-1*). Connect the battery, set switch S1 to *PNP* (the *down* position) and press switch S2. LED1 will light up, indicating the unit is set up for PNP transistors. Go to the *NPN* setting (S1 *up*), and LED2 will light when S2 is activated.

Now, insert a known *good* NPN transistor in the appropriate socket and adjust potentiometer R5 to full clockwise. Hit the test button (S2) and you should see a meter reading of about half-full scale. It might be slightly higher or lower, depending on the exact transistor under test, but an indication somewhere within this ball park signifies a properly operating device.

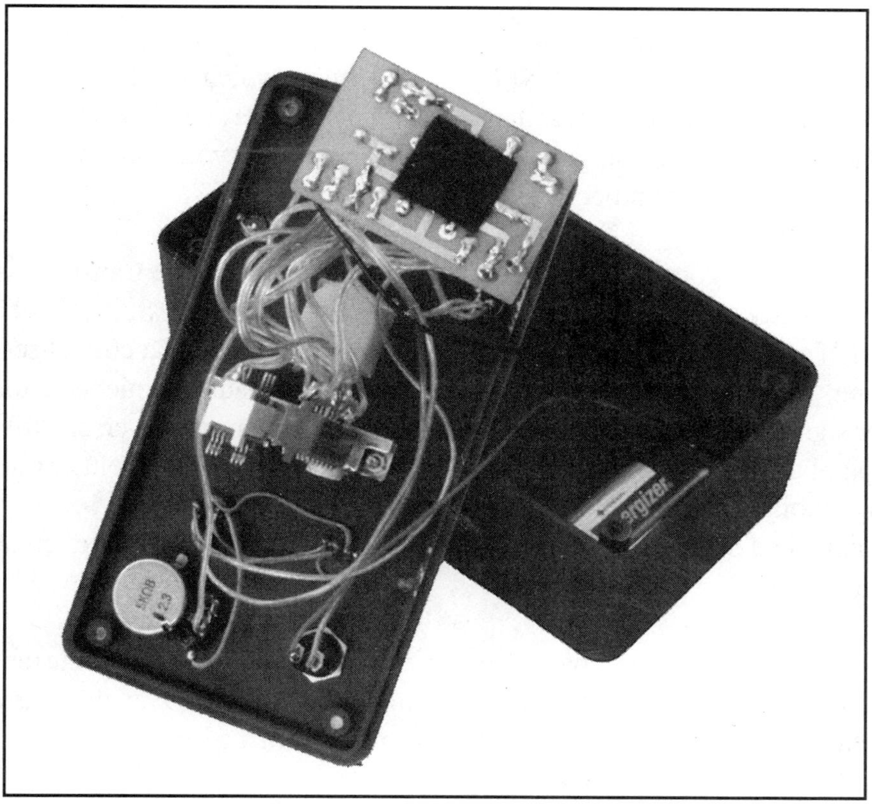

Photo 5-1. Internal shot of the completed transistor tester.

Next, rotate R5 counterclockwise; the meter should show a decrease in voltage. This indicates that the transistor's base is controlling the emitter-to-collector transfer. With the recommended value for R5, the meter should fall back to about one third full scale with R5 at full counterclockwise.

For the leakage test, press switch S3, and a slight rise in voltage should be seen. A large voltage rise is bad news. This test can also be used to identify transistors that are still functional but possess excessive leakage. It is best done with R5 set to the full clockwise position.

If you have a known bad transistor available, try it out. A shorted unit will *peg* the meter, while no meter movement indicates an open device, if you know the type of transistor you are testing. If not, as you will see in a moment, there may be other problems. Read on!

Two additional tester functions are used to recognize the type of device being tested (either PNP or NPN) and determine which leads are which. With an unknown transistor in one of the sockets, set switch S1 for PNP, and press *Test*. If you get a normal meter reading, you have the right lead designations and polarity setting.

If the meter goes to full scale, either S1 is wrong, or the transistor is incorrectly plugged into the socket, or the device is bad. So, set S1 to NPN and try again. A normal meter reading will indicate a correct setting, while another full scale will either confirm a faulty semiconductor or signify still erroneous lead connections. A little trial-and-error shifting of the leads and switch S1 will soon tell the tale. You will either find a combination that works, or you will indeed have a dirty, rotten, lousy, faulty, bad ol' transistor that has earned a one-way ticket to the circular file.

Incidentally, the 9-volt battery doesn't apply enough voltage or current to cause reverse-polarity damage in most transistors. Over the many years I've used my unit, I have yet to end up with a casualty.

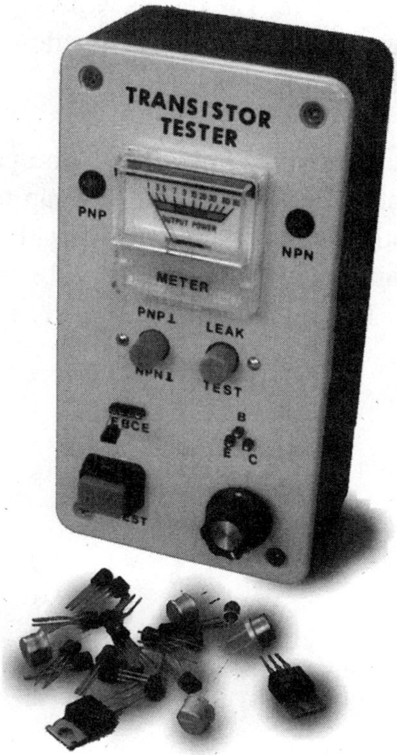

Photo 5-2. External view of the completed transistor tester.

Conclusion

You've done it! Now, when you purchase a double handful of surplus or unmarked transistors, they will be no match for you. With this trusty tester in hand, you'll know what they are in the blink of an eye. Well, almost that quickly. In all seriousness, though, while not a complicated device, this unit will come in quite handy (*Photo 5-2*). Aside from testing new or stock devices, it can determine if a transistor you remove from a problematic circuit really is the culprit.

For that matter, the addition of a set of test leads will allow for the examination of devices *in circuit*. The results will vary, but you should be able to tell if a suspected semiconductor is functioning properly. This, of course, is a possible modification.

So, if testing transistors is your thing, this is the project for you. A handful of components, a little time, and you can test to your heart's content, or at least until you run out of transistors. Either way, with this gem at your side, those polarized junctions will never again have the proverbial last laugh. That, at least in my mind, is awfully good news.

Chapter 6
An All-Purpose Tester

Introduction

Over the years, I have run into dozens of designs for integrated circuit (IC) testers. A few were as simple as a solderless breadboard arrangement allowing for the construction of circuits that tested LM555s or LM741s, or other chips. Some employed microprocessors, EPROMs and digital displays resulting in very complex devices. However, it been my experience that the middle-of-the-road variety is the most useful and cost efficient.

Again, a wide spectrum of configurations has been seen, but for the most part, many IC testers have one thing in common: they run a continuity check on the various positively doped/negatively doped (P/N) junctions within the integrated circuit.

That's how we are going to do ours. What I'm presenting here is a compilation of the best features of the many designs I've encountered; bits and pieces that stuck in the old gray matter. The test unit we're going to build provides an inexpensive yet dependable method of checking ICs, both digital and linear. Once completed, this IC tester reads the internal status of the device under test, and displays the status with light-emitting diodes (LEDs). When functioning properly, each type of integrated circuit exhibits a unique pattern of illuminated LEDs, and this is the key behind our tester. So, let's take a closer look at how this system works.

INTEGRATED CIRCUIT TESTER PARTS LIST

SEMICONDUCTORS

U1	LM7805T 5 Volt Positive Regulator
Q1-24	2N3904 NPN Transistors
LED1-25	T-1 3/4 Jumbo Light Emitting Diodes (Assort. Colors)
LED26, 27	T-1 Light Emitting Diodes (1-Red, 1-Green)

RESISTORS

R1-12, 37-48	1,000 Ohm 1/4 Watt Resistors
R13-36	10,000 Ohm 1/4 Watt Resistors
R49	220 Ohm 1/2 Watt Resistor
R50, 51	220 Ohm 1/4 Watt Resistor

CAPACITORS

C1	470 Microfarad Electrolytic Capacitor
C2	1 Microfarad Electrolytic Capacitor

SWITCHES

S1, 2	SP12T Rotary Switches
S3	DPDT Slide Switch
S4	SPDT Slide Switch
S5	SPST Push Button Switch
S6	Momentary Contact Push Button Switch

MISCELLANEOUS

Phono Jack for Power, PCB Materials, Solder, Hook-Up Wire, Knobs, Dry Transfer Lettering, Paint, Case (9" x 5 1/2" x 2 1/4"- All Electronics #MB-40), 24 Pin ZIF (Zero Insertion Force) IC Socket, Etc.

Theory

A close examination of *Figure 6-1*, the schematic, quickly reveals 24 identical circuits connected to a 24-pin zero-insertion force (ZIF) IC socket. Each is comprised of a transistor, LED and two resistors; and if you were paying attention in Chapter 4, yes, each is a logic probe in its simplest configuration. With these probes, a high signal lights the LED. Thus, when the P/N junctions are functioning as anticipated, a positive voltage applied to the proper IC pin (VCC) will send certain (or sometimes all) pins high. The probes then detect those highs and light their respective diodes.

As a quick review, remember that the transistor can act as a switch, where a potential on the base allows current flow from the emitter to the collector. The amount of flow is proportional to the level of the base potential. Consequently, the higher the base voltage, the higher the emitter-to-collector transfer. This can approach the supply voltage, although it will never completely reach it as there is always some loss in any P/N junction. A transistor, of course, is such a junction. Also, the number of junctions within the integrated circuit will affect the amount of voltage that reaches the probe. Naturally, this is due to the various IC junctions absorbing small portions of the full voltage.

It follows that the brightness of each individual LED will vary depending on the complexity of the circuit under test. If the applied voltage has to travel through a bunch of junctions, then the LED is not going to glow as dazzlingly as if it only went through one or two. However, the presence of any glow at all is usually a good sign the IC is working, as a bad P/N junction normally blocks most of the current; at least enough to keep an LED from reaching threshold and lighting.

So, once a particular IC pattern or signature is established, all like types can be tested against the signature. Each properly operating device will display the same pattern of illuminated LEDs. As for the specific signatures, this will be covered in greater detail later.

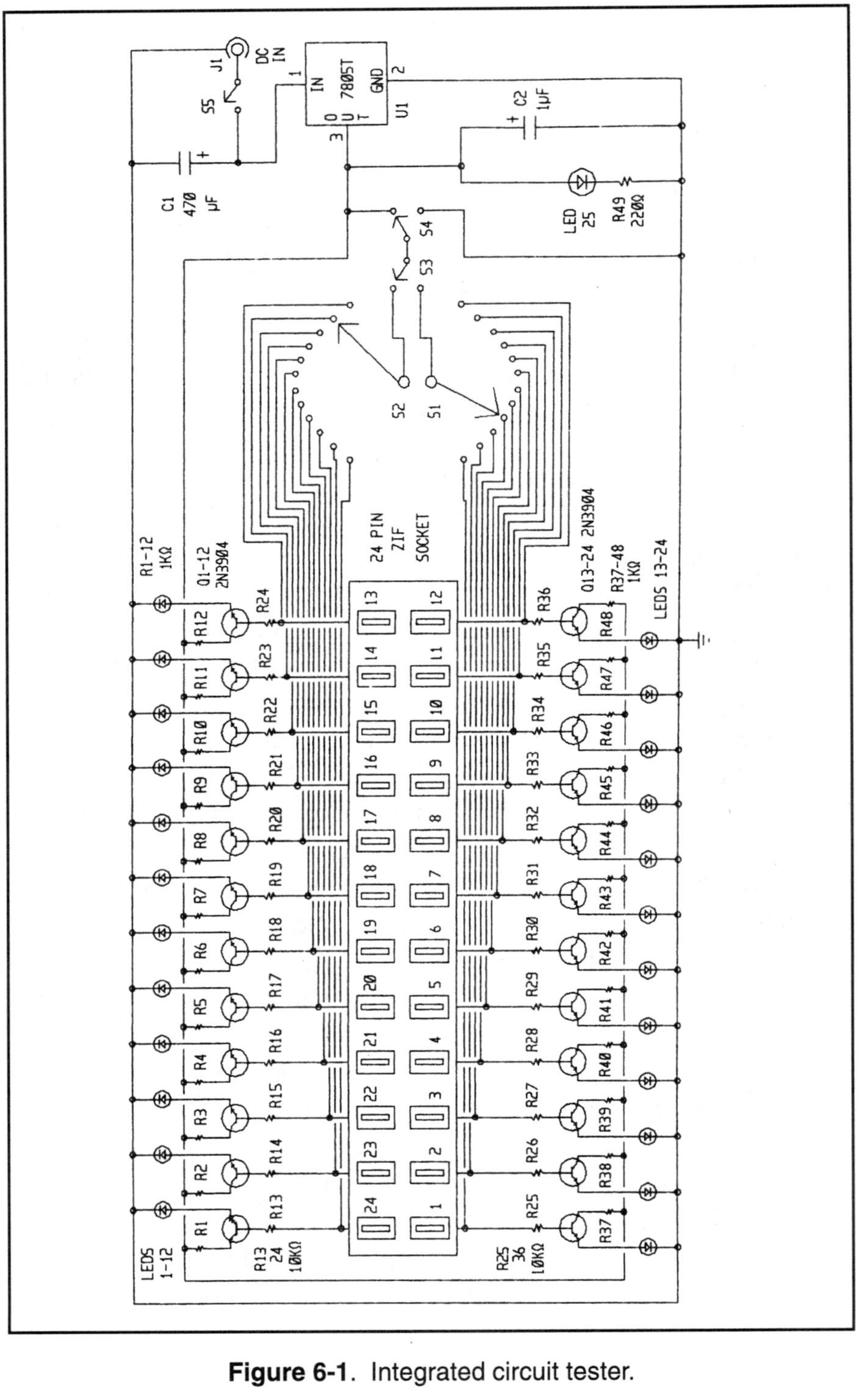

Figure 6-1. Integrated circuit tester.

Back at the ranch (*Figure 6-1*), you will notice that two SP12T rotary switches are employed. Each switch connects one throw (T) to one pin of the ZIF socket. Those connections are made between the pins and the base-dropping resistors (R13-36). This allow for the assignment of VCC to the proper device pin; say pin 14 for a 74LS04 hex inverter.

As for ground, it too can be properly assigned via these switches (S1 and S2), but not at the same time VCC is being applied. That's okay; only one potential needs to be connected at any given time. This will be covered in the operations section.

Switch S4 decides whether the rotary switches receive positive voltage or ground, while switch S3A determines which rotary gets the signals. S3B, the second section of the DPDT slide switch, is used to visually indicate which rotary switch is in use. This is done through LEDs 27 and 28.

As a sidebar, here, these two diodes are passive in nature and serve only as a convenience; they can be left out. However since I've already employed 25 LEDs, and someone remarked this project looks like a Christmas tree anyway, I figured a few more LEDs wouldn't hurt. Besides, it is nice to know at a glance which rotary switch is active.

Integrated circuit U1 serves as the nucleus of a regulated power supply. I chose a 7805 because, to my knowledge, nothing this tester can test will be hurt by 5 volts. Since a DC wall-type transformer is used for input power, a transformer/rectifier section is not necessary.

Capacitor C1 provides initial filtering, while capacitor C2 prevents U1 oscillation and filters transients. Resistors R49 and R50 provide dropping for LEDs 26 and 27/28, respectively. Switch S5 is for main power, and S6 is the test button, applying +5 volts to the main board when pressed. S6 also lights LED25 (the *test* indicator).

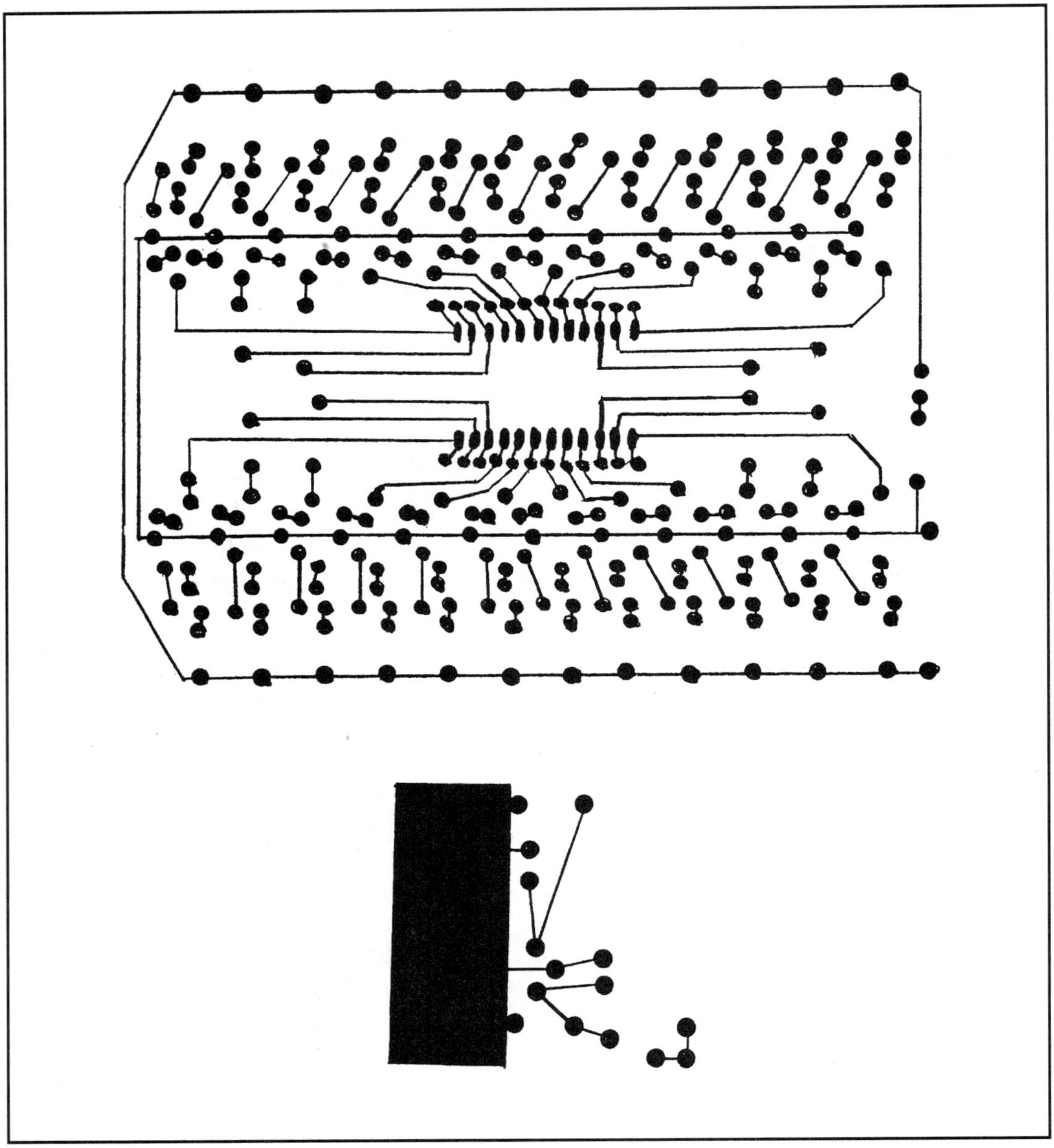

Figure 6-2. IC tester foil pattern and power supply.

So, there you have the IC tester's theory in all its glory. With that under our belts, let's move on to constructing this device. You know, the fun part.

Construction

Here, as usual, we have to make a decision between using point-to-point wiring or a printed circuit board (PCB). Also, as usual, I went with two PCBs; the foil patterns are shown in *Figure 6-2*. Parts placement is handled by *Figure 6-3*. A point-to-point layout will probably follow the PCB arrangement fairly closely, so these instructions should be acceptable for both schemes. Just be careful to avoid solder bridges if you go the perf board route.

For the main board, first install the jumpers. There are 32 of the little suckers. Next comes the standard 24-pin IC socket. This is employed to protect the expensive ZIF socket you bought. Now, place all 97 resistors. Next come the LEDs, and these require a little care. You want all of the 24-pin indicators to line up in two nice, straight rows (one on each side of the IC socket), and be the same height from the PCB surface.

The first part I did by eyeballing, which works quite well. For the second requirement, virtually every T-1 3/4 jumbo-style LED I've ever worked with has a slight widening of the leads beginning just below the diode's base. When this enlarged area is mounted flush with the board's surface, the top of the LED will be exactly 1/2" off the board. This takes care of the height problem. Last are the transistors, and again, they need to be installed with some care. Aside from proper orientation, you want the top of the TO-92 packages to come to just about the base of the LEDs. This allows the diodes to protrude sufficiently through the case without the transistors getting in the way. This completes the main board, as seen in *Photo 6-1*. Well, almost.

As for the power supply board (*Figure 6-3b*), this is the easy part. It only holds four components, and there are no special height or positioning requirements. Just make sure the leads fit the holes (though this might be construed as stating the obvious). The PCB recommended provides ample excess copper surface to handle heat sinking for the regulator. Be sure to leave the copper on both sides and use heat sink compound on the 7805.

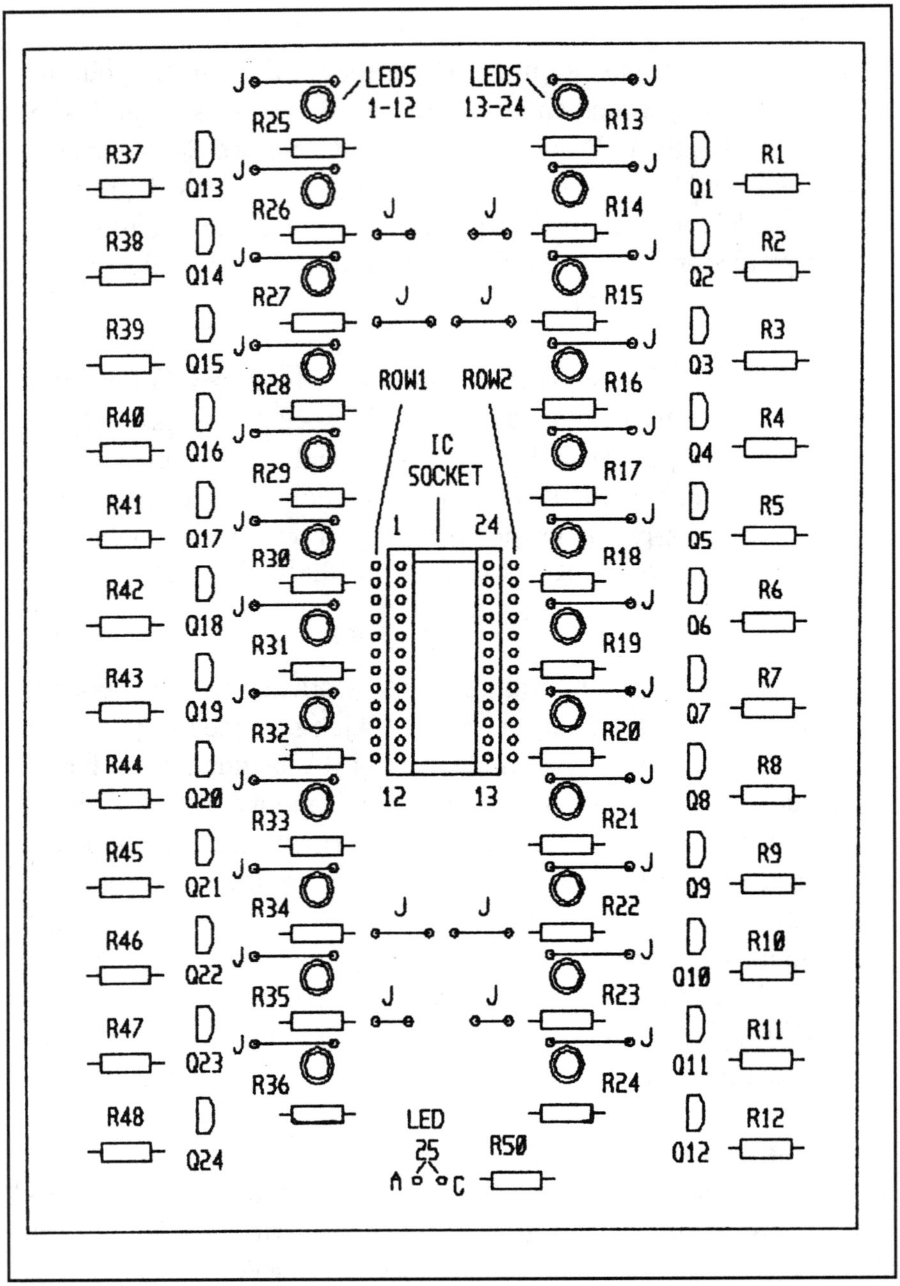

Figure 6-3a. Parts placement.

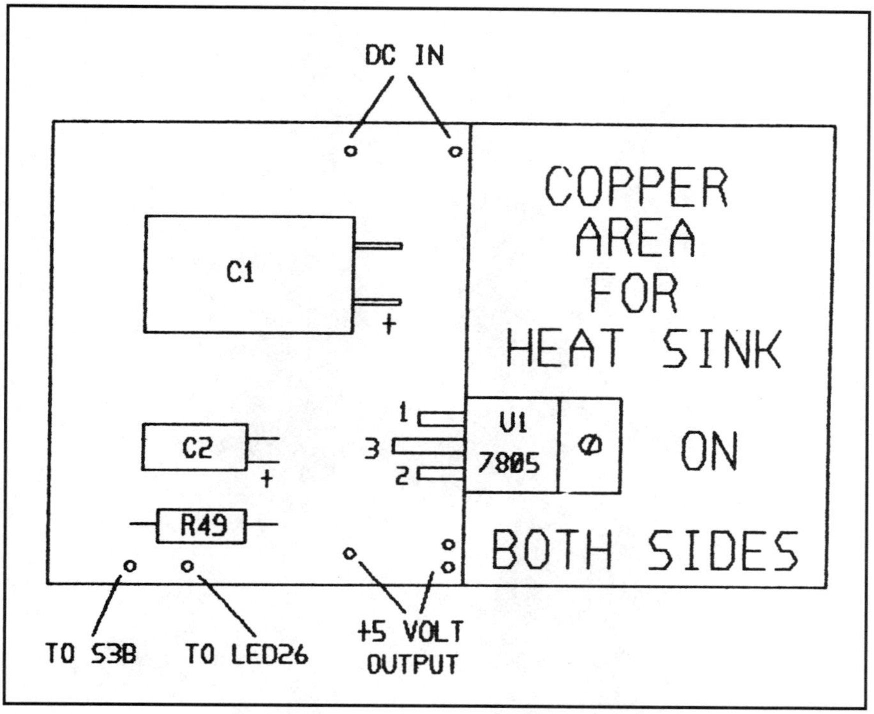

Figure 6-3b. Power supply parts placement.

Returning to the main board, it's now time to connect the two rotary switches (S1 and S2). Using sufficient lengths of hookup wire, solder a lead from each switch's throw terminal to each pin's connection point on the board. S1 is for the left side of the socket while S2 handles the right side.

It's also best to gather the wire bundle from each switch and secure it with a cable tie. This keeps the project orderly and helps prevent strain on the leads.

The poles of the rotary switches go to the throw positions of switch S3A, and its pole goes to the pole of switch S4. One throw of S4 is tied to +5 volts, while the other is grounded. S3B's pole is connected to the positive line through resistor R50, with the anodes of LEDs 27 and 28 going to the throws.

Photo 6-1. The stuffed PCB, showing the LEDs, transistors and other components in place.

I hope all of this doesn't throw you. I know, switches can get really confusing at times. Just work carefully and keep *Figure 6-1* handy.

At this point, let's turn our attention to the enclosure. I packed this beauty in a surplus case I found measuring 9" (230 mm.) by 5-1/2" (140 mm.) by 1-1/2" (38 mm.). Any case close to those dimensions should do the trick. Of primary importance is to leave one surface with plenty of open area for the panel components.

Figure 6-4 provides a template for the prototype that allows ample space for everything. Using drills, files and a small hobby saw, make all of the

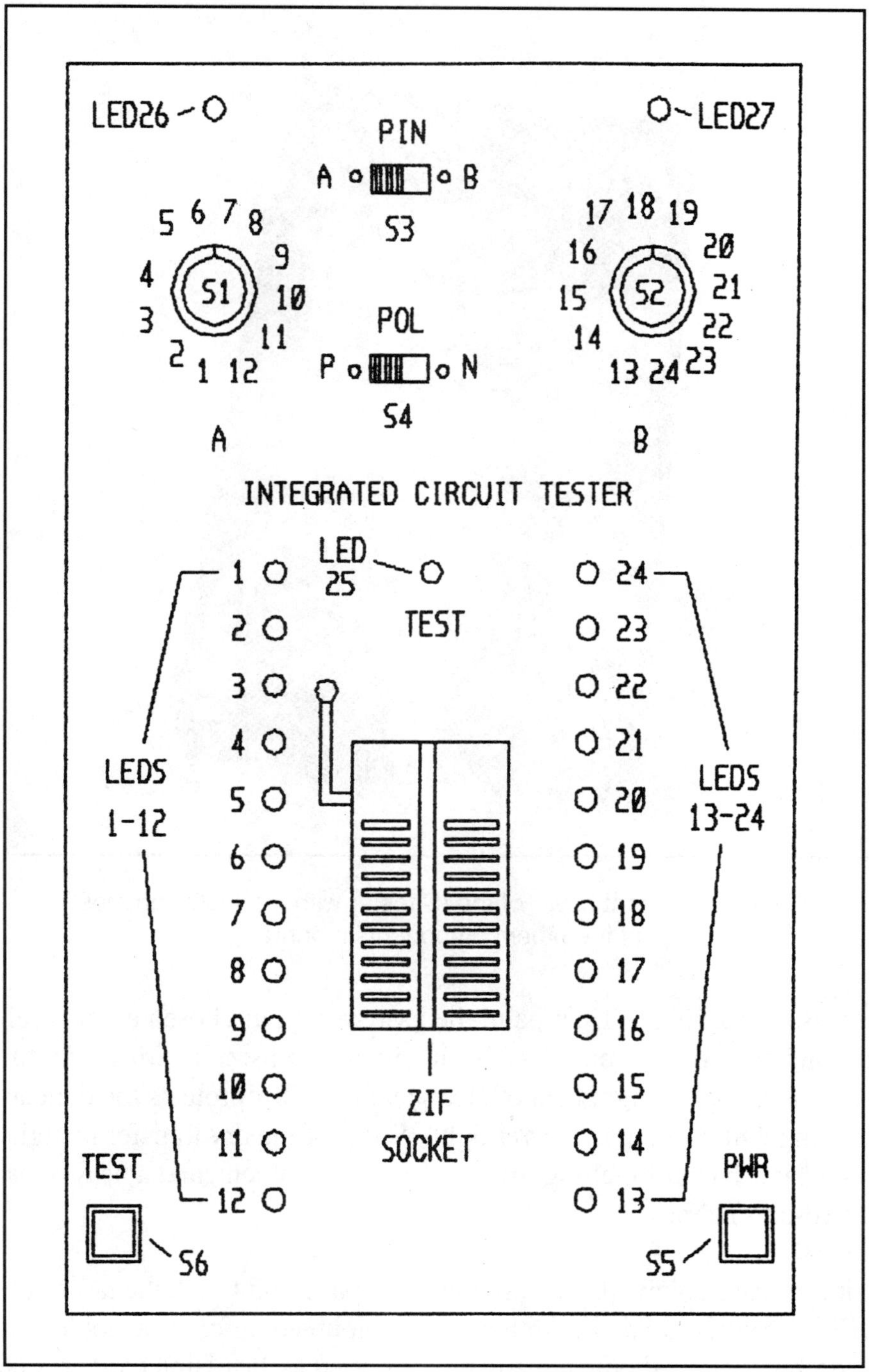

Figure 6-4. Panel template.

Photo 6-2. Front cover of the IC tester with the PCB and most of the other components mounted.

necessary openings. Then paint the case with your chosen colors (yellow and blue for my unit) and label it with dry-transfer lettering. For the last touch, apply a light coat of clear enamel. This protects the delicate lettering, but do keep this layer light. Some of the dry transfer material has a nasty habit of melting from the enamel's solvent, and a heavy coat can result in a mess.

With the case completed, install the panel parts and finish the last of the wiring. Switches S5 and S6 have to be mounted before you solder the leads, but the rest should already be connected to the main board or each other, via hookup leads.

As seen in *Photo 6-2*, most of the circuit fits on the top cover of the case; this makes it easy to handle. As *Figure 6-3* illustrates, the power supply mounts on the bottom half with bolts and spacers, and is connected with leads to the switches, LEDs and main board. A cable tie keeps everything neat.

Originally, I had planned to anchor the main board with screws and standoffs. As it turned out, I found that the two rows of LEDs fit tightly enough that a little epoxy on the corner diodes holds the board quite well. The epoxy, of course, is applied from the inside to the base of the LEDs. It wouldn't look right on the outside.

If all is in place, the case can be closed up since nothing inside needs to be adjusted. Install the knobs and ZIF socket, and your tester is ready for testing (*Photo 6-3*).

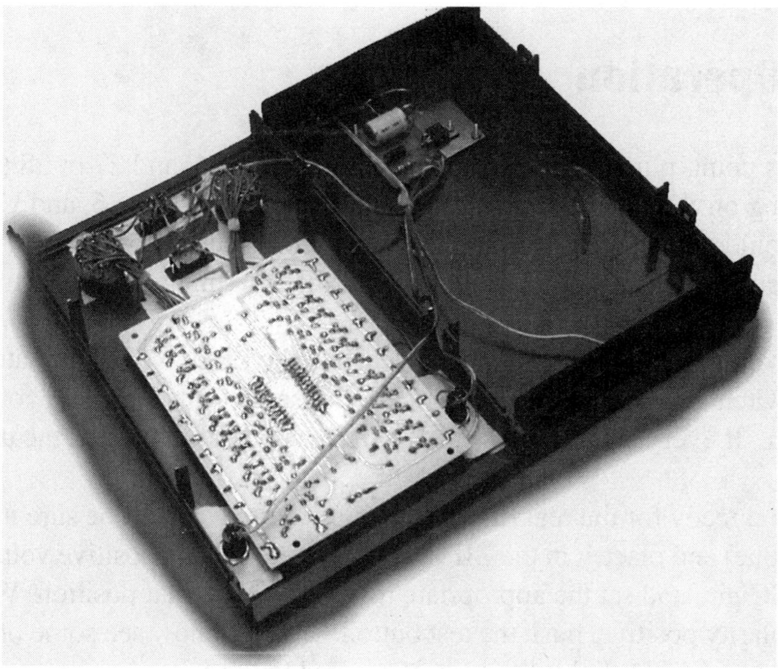

Photo 6-3. Both halves of the IC tester, in the case. The bottom half holds only the power supply.

Device	Sock.Pos.	LEDs Lit	LEDs Dark
7400	top	all	none
7404	top	all	none
7447	top	all	none
7490	top	1,2,3,5,6,7,18-22,24	4,23
74190/193	top	all	none
4001	top	3,4,20,21,24	1,2,5,6,7,18,19,22,23
4049	top	1,2,4,6,18,20,23	3,5,7,17,19
4069	top	2,4,6,18,20,22	1,3,5,7,19,21,23
4093	top	3,4,20,21,24	1,2,5,6,7,18,19,22,23
324	top	all	none
339	top	all	none
555	top	1,2,3,4,21,23,24	22
556	top	1,2,4,5,6,7,18,19,20,21,23,24	2,22
741	top	1,4,21,22,23	2,3,24
1458	top	1,2,3,4,21,23,24	22 out/dim

Table 6-1. Results from the IC tester.

Testing and Operation

At this point, plug in the wall transformer; LEDs 26, and 27 or 28 (depending on S3B's setting) should light. Push test switch S6, and LED 25 should glow. With polarity (POL) switch S4 set to positive (P), the LED corresponding to the active rotary switch setting will also light.

Next, rotate the active rotary switch; each of the LEDs will illuminate as the switch reaches their position. Do the same with the other rotary switch. If everything checks out, you have successfully tested the unit.

Are you ready for the real fun? Find an integrated circuit (be sure it's a good one) and place it in the ZIF socket. Determine the positive voltage or VCC pin, and set the appropriate rotary switch to that position. With the polarity positive, push the test button. You will now see some or all of the panel LEDs light; the ones connected to the IC.

For example, let's say we tested a 74LS90. It doesn't matter how you place the IC in the socket, as all pins are wired to identical circuitry; but for this scenario position the 74LS90 at the top. In this way, the left row matches the actual pin numbers; but, of course, the right row doesn't. Now, press *test* and note which LEDs glow. I found that for every *good* 7490, LS90, etc., pins 1, 2, 3, 5, 6, 7, 18, 19, 20, 21, 22 and 24 illuminate, while pins 4 and 23 do not. For this IC placement, pins 18-24 would equate with the actual pin numbers of 8-14. Thus, the given pattern is the 7490 signature.

In the future, when you test a 7490 or one of its series counterparts, the same LEDs will light up. If not, the device can be considered bad.

Naturally, you will need to test as many ICs as you have handy and develop a list for each. Also, as you acquire new devices, it is a good idea to test them and add the results to your list. As a starting point, I have included *Table 6-1*. This contains the results I obtained from some of the more commonly used ICs.

Simple, eh? The only real trick is knowing which pin is VCC, but this is usually not difficult information to obtain, if you do not already have it.

As for a few guidelines, with the digital ICs, I found that the 7400s or transistor-transistor logic (TTLs) tend to light all LEDs involved. The only exception is when a *not-connected* (NC) pin is encountered; but this goes for all NC pins with any type of integrated circuit.

The complementary metal oxide semiconductors (CMOS) or 4000 series will only light diodes connected to the output pins. Thus, you will usually have a substantial number of dark LEDs. Also, the diode linked to the VCC pin will always be lit, for obvious reasons.

A second test can be done by changing the polarity of the VCC pin to negative or ground. This one normally has no effect on the digital devices other than to turn them all off, but it can sometimes be helpful for determining correct signatures for analog or linear ICs.

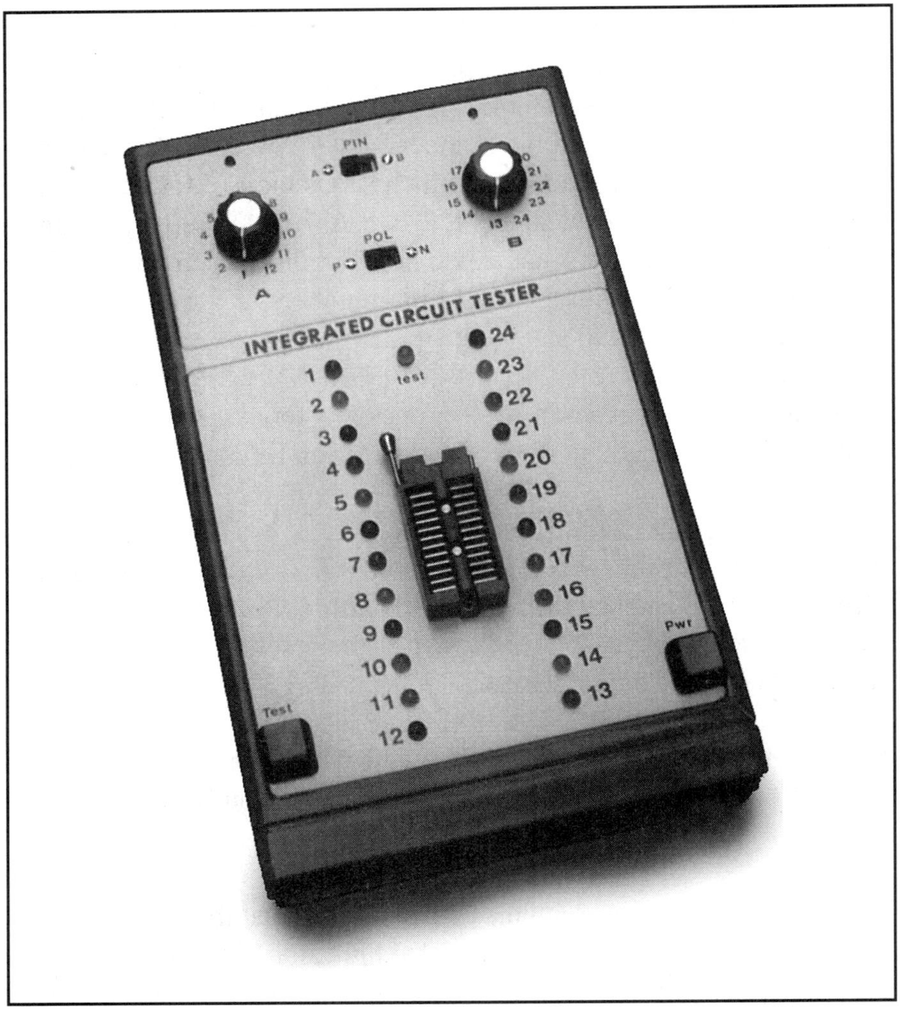

Photo 6-4. The completed IC tester, ready for use.

As stated before, a listing of each linear integrated circuit needs to be made by testing good chips. In doing so, try reverse polarity on each and note the results. Most will not respond, but for the ones that do, this will provide additional test data.

There you have it. Not a difficult device at all. However, it's a device that will come in very handy the next time you buy a double handful of those cheap surplus chips.

Conclusion

I've been in the field of hobby electronics for, shall we say, many moons, so I've seen tubes and transistors, and finally integrated circuits. ICs are miraculous devices that save money, time and space, but they do have one drawback: you can't get inside to check individual components when a circuit doesn't function properly. Of course, you could replace the entire chip, but this is not always a convenient or plausible option.

So, the next best thing is to be able to check the IC itself. This process can definitely prevent huge doses of frustration, especially if you don't regularly use IC sockets. Naturally, it goes without saying that this has been the objective of this chapter and the resulting project. So, with the IC tester in hand (*Photo 6-4*), you will be able to test those otherwise mysterious little devices we all know and love.

Knock 'em dead, Ace.

Chapter 7
A Portable Digital Capacitance Meter

Introduction

If you are an electronics hobbyist, you've probably purchased surplus capacitors in bulk. Everybody does. It's an excellent way to stock up the parts box. However, one drawback is that often perfectly good capacitors are marked with house numbers or not labelled at all. Unless you have a capacitance meter, you are left to guess the values, usually by size, shape, etc. Not a very good approach, to say the least.

Enter our digital capacitance meter. In a small, hand-held package, you possess a device that confirms both capacitor operational status and actual value. It also uses its own self-contained batteries for convenience and practicality. Too frequently, the markings on so-called *prime-quality* capacitors are less than accurate, especially for the ceramic disks; so the capacitance meter will be very valuable in this realm.

I know you will enjoy having this device on your workbench. It will come in handy so often, you will probably wonder how you survived without it.

Theory

For this project, we are going to take a different approach. Since our meter uses a digital voltmeter for a display, and we have already built ICL7107 meters for the prototype lab and power supply (don't

CAPACITANCE METER PARTS LIST

SEMICONDUCTORS

U1	74HC132 Quad 2-Input NAND Gate Schmidt Trigger
U2	7805L 100 Milliamp Positive 5 Volt Regulator
D1-3	1N914/1N4148 Signal Diodes
DVM	3-1/2 Digit 7106 LCD Panel Meter Module (See Text)

RESISTORS

R1	200,000 Ohm PCB Potentiometer
R2	47,000 Ohm 1/4 Watt Resistor
R3	2,000 Ohm PCB Potentiometer
R4	680 Ohm 1/4 Watt Resistor
R5, 10	1,000,000 Ohm 1/4 Watt Resistor
R6	100,000 Ohm PCB Potentiometer
R7	5,100,000 Ohm 1/4 Watt Resistor
R8	10,000 Ohm 1/4 Watt Resistor
R9	120,000 Ohm 1/4 Watt Resistor
R11	100,000 Ohm 1/4 Watt Resistor

CAPACITORS

C1	390 Picofarad Monolithic or Mica Capacitor
C2, 3	1 or 1.5 Microfarad Tantalum Capacitors
C4	10 Microfarad Aluminum Electrolytic Capacitor
C5, 6, 7	0.1 Microfarad Monolithic Capacitors

OTHER COMPONENTS

S1	SPDT Slide Switch
S2	DPDT Slide Switch
S3	SPDT Slide Switch
J1, 2	Banana Jacks and Header Jack

MISCELLANEOUS

PCB, or PCB Materials, Solder, Battery Connector, Hook-Up Wire, Portable Meter Case, Paint, Dry Transfer Lettering, 9 Volt Battery, 14 Pin IC Socket, Hardware, Etc.

fret; you'll get another shot at the 7107 in Chapter 9), I elected to use a factory-assembled 7106 (LCD version) voltmeter. These compact panel meters, handled by a number of distributors, will run for 200 hours or so off a 9-volt alkaline battery. Best of all, they are cheap, usually under fifteen dollars or so.

As with the 7107 design, the 7106 contains the analog-to-digital (A/D) converter, clock, reference, decoders and liquid crystal display (LCD) drivers, all on a single 40-pin chip. External circuitry is virtually identical to its cousin, except for the power supply.

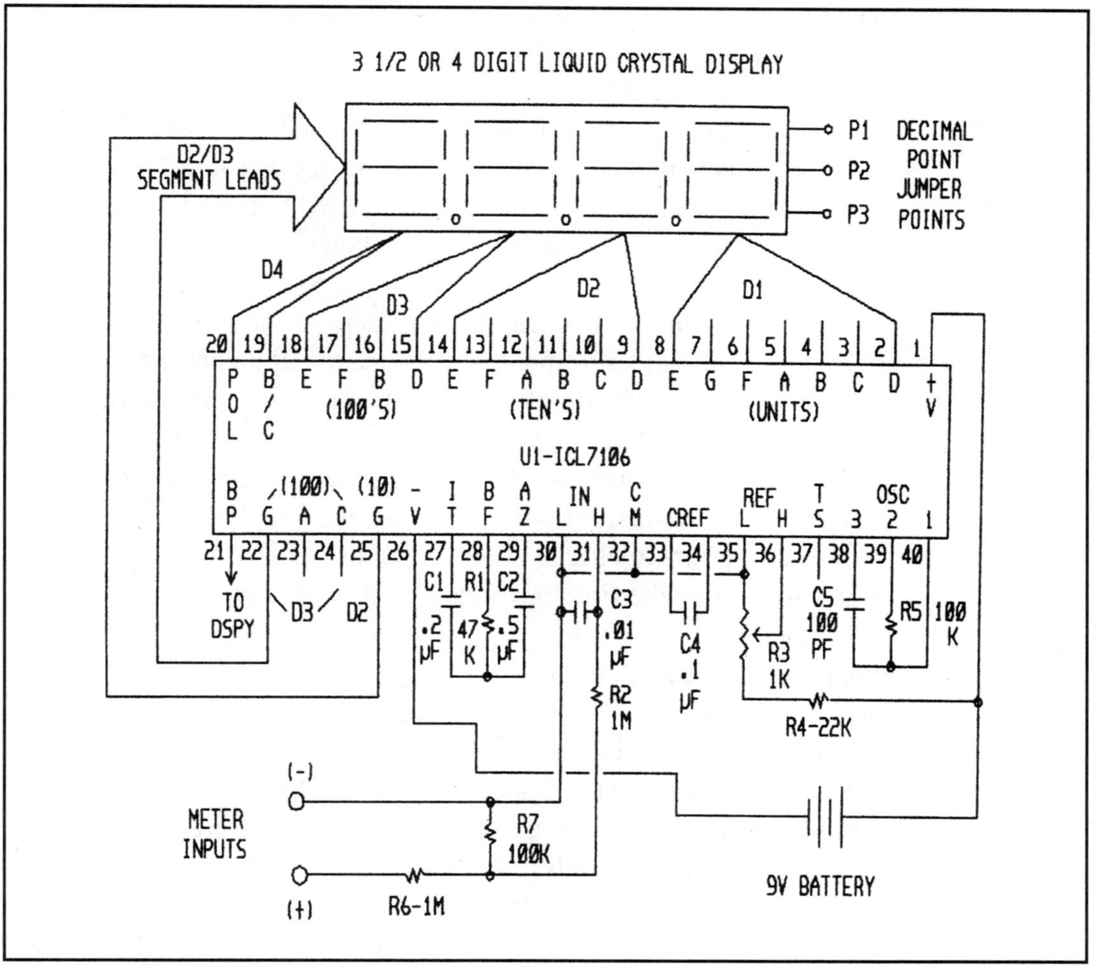

Figure 7-1. ICL7106 LCD meter schematic with voltage divider for 2-volt readings.

With the 7106, the positive and negative voltages remain at pins 1 and 26, respectively, but pin 21 is used for the LCD display *backplane* drive instead of ground. This allows for battery operation and eliminates the need for a split (+/-) power supply.

The remaining external components have stats virtually identical to the 7107. Since most of these panel meters arrive as a *basic* system, they read a maximum of 200 millivolts full scale. A simple 1-megohm/100-kilohm voltage divider allows the meter to perceive 2 volts full scale (see *Figure 7-1*). This is what we will need to read signals from the second half or capacitance-measuring section of the design. The two working together furnish an accurate value measurement that is conveniently displayed digitally with the voltmeter.

A quick note here: you will need two separate 9-volt batteries for this system. This is due to the fact that the voltmeter cannot read its own voltage; thus, its power supply cannot be used to power other circuits. Attempting to do so will result in a conflict.

As for the second half of the design, there are a number of ways to obtain the value measurement. Appropriate circuits range from quite simple to downright complex. For example, the measuring can be done with an oscillator and frequency-to-voltage converter, in which the capacitor under test is part of the oscillator's timing circuit. Another system would start with the oscillator, use NOR gates, transistors and inverters to charge other capacitors and sample-and-hold circuitry to display the findings on a digital voltmeter.

So, as can be seen, this project may be as simple or as difficult as you desire. While more complex configurations usually afford greater measuring range, it has been my experience that accuracy is seldom better than with a simple system. Thus, I have selected a single integrated circuit (IC) design that furnishes excellent precision and will measure from 2 picofarads to 2 microfarads. This range covers all but the large electrolytic capacitors, which are almost always well marked even when you are dealing with surplus sources. Frankly, you will encounter few situations in which this range doesn't cover your needs.

Let's look at the circuit. The *different* part is that it is built from a kit, specifically the CA-1 *Capacitance Adapter for DVM* kit from Electronic Rainbow (see the Appendix for a source list). A number of distributors carry Rainbow's line, or parts be ordered directly from the manufacturer.

The advantage to going this route is that, as a kit, the CA-1 contains all components and an already etched and drilled printed circuit board (PCB). Additionally, the CA-1 interfaces directly with any digital voltmeter set for the 2-volt range. Remember, this is our digital panel meter's setting.

A final bit of good news is that the last time I looked, the CA-1 kit rang in at under fifteen dollars; thus, the whole works shouldn't cost over thirty dollars. I hate to sound crass regarding the money angle, but where hobbies are involved, most of us have to watch the pennies. At least a little anyway.

As stated before, the circuit is based around a single 74HC132 quad 2-input NAND gate Schmitt trigger. Actually, this system will work with a 74HCT132, a 74LS132 or, with a little less reliability, a 7400. All are NAND gates with identical pinouts.

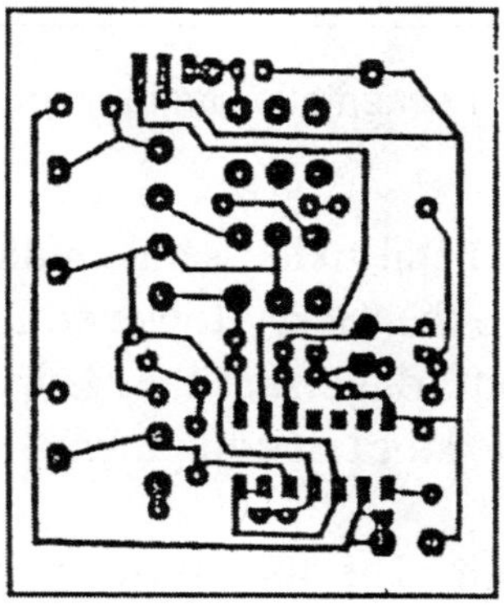

Figure 7-2. Rainbow kit foil pattern.

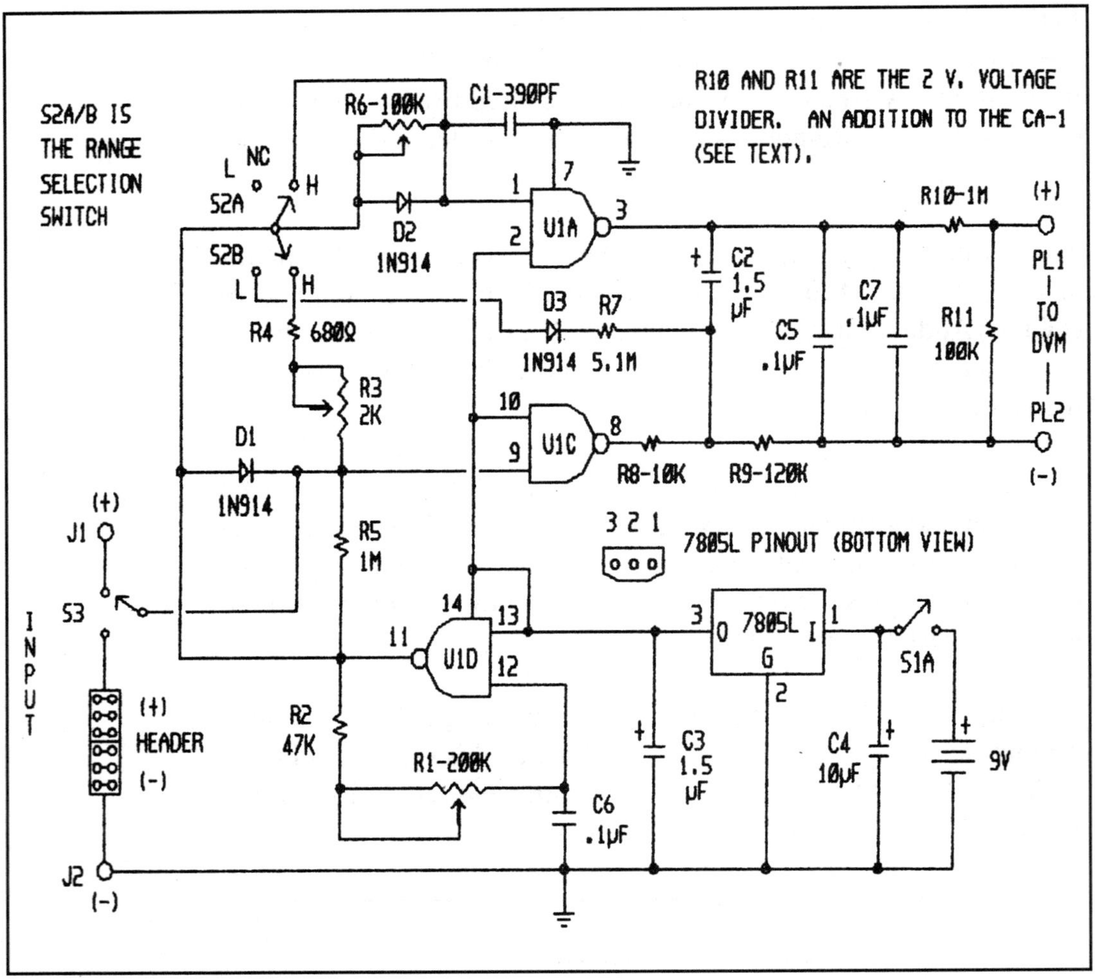

Figure 7-3. CA-1 schematic diagram.

On the complementary metal oxide semiconductor (CMOS) side, a 4093 or 4011 will also do the trick. However, I recommend using the 4093 NAND with Schmitt triggers, as they help the device's stability. This is an important factor for test equipment.

A word of caution, though. The 4000 series pinout is different for the *C* gate, with pin 10 as the output instead of pin 8. It follows that the inputs are pins 8 and 9 for the CMOS devices, as opposed to 9 and 10 for the 7400 series.

This information is all for the sake of reference, as the kit will come with a 74HC132 (or at least the one I bought did). I made some minor modifications involving the switches, but other than that, the unit went together per the instructions. (*Figure 7-2*)

Since the switches are panel-mounted, I had to take them off the PCB and connect them with hookup wire. This will be covered in more detail later. This is essentially a minor change that can be done with very little trouble.

Back to the circuit. This design is another example of a clever alternative used for a standard logic IC. Here, one of the Schmitt trigger gates is configured as an oscillator while two more comprise a pulse-width comparator.

Looking at *Figure 7-3*, the CA-1 schematic, you will note that one input of all three gates (the fourth gate is unused) is connected to the positive 5-volt rail. According to the truth table for NAND gates, this will result in a positive or *high* output unless the remaining input (or inputs) also goes high.

For the oscillator section, U1D employs capacitor C6 and resistors R1 and R2 in an R/C tank circuit. The charging and discharging of C6, controlled by potentiometer R1, permits the second input (pin 12) to go high when the 74HC132 trigger voltage is reached (about 2-1/2 volts).

With both inputs high, U1D's output goes low, allowing C6 to discharge. When the capacitor's charge drops to about a volt and a half, pin 12 goes low and the whole cycle starts again. I'm certain you recognize this as oscillation. Anyway, as for the comparator section, gates U1A and U1C do the honors. The unused inputs (pins 1 and 9) follow the action of pin 11; thus, square wave outputs are present at pins 3 and 8. This is accomplished in a fashion similar to the pattern used for U1D.

The outputs furnish both legs for the external digital voltmeter connection, with pin 3 providing the positive side and pin 8 the negative. Potentiometer R6 is adjusted so that without a test capacitor connected between J1 and J2, the square wave at both pins is the same. This results in a display reading of *000*. The comparison action takes place when a capacitor is attached to the input points (J1/J2). The capacitor under test becomes part of the U1D oscillator system. This, in turn, increases the length of time the U1C leg is *LOW*.

Pin 8 is now low longer than it is high. However, U1A remains high longer than it does low, and the difference (pulse width) sets up a positive voltage proportional to the test capacitor's value. The voltage is then filtered by capacitors C2, C5 and C7 and resistors R8 and R9 to provide a stable DC output. It is then sent to the digital voltmeter.

So far, we have only addressed the CA-1 in its *low* range. For the *high* setting, additional resistance is introduced to the U1D circuit, which further slows down the *low* timing. Again, the comparator action of the U1A/U1C circuitry adjusts the voltage output to represent a higher capacitance value.

The last section of the CA-1 is the power supply. As with all good test equipment, this device utilizes a voltage regulator to ensure stable operation. In this case, the 7805L is chosen for its smaller size, and is employed in the usual linear fashion. I guess I lied when I said the CA-1 is a single IC configuration. The 7805L makes two.

So, that's that. Once the CA-1 is properly calibrated (which will be covered in the construction section), the unit can read unknown capacitance from 2 picofarads to 2 microfarads, using only two range settings.

Construction

Since the digital voltmeter is already constructed, and the CA-1 comes with a PCB, the decision to use point-to-point or PCB is made for us.

However, we need to make a few modifications to both units. Let us look first at the voltmeter.

The unit I ordered came from JDR Microdevices, marked only as *C+C, PM-128 Digital Panel Meter*. The rather sparse instruction sheet describes how to set up the meter to read various voltages. However, I found that it was better to simply incorporate (externally) the previously-mentioned voltage divider.

To set the decimal point (DP) to the fourth place, remove five small screws from the backside. This allows the assembly to separate. Beware, though; connections to the LCD panel are made with conductive rubber strips. Leave them in the front part of the case.

The fourth DP position is labelled *P3*. The display is accomplished by shorting the two board points marked *P3* and *ON* with a piece of jumper wire soldered to the inside surface of the board.

Next comes the power. The two board points are marked with the battery symbol. The longer line indicates *positive* voltage. I used red and black lead wire and again soldered them to the inside surface of the PCB.

Now you need to do the same thing with the two input points. Use lead wires of whatever colors you want for the positive and negative connections. These will later go to the voltage divider and CA-1 outputs. You could also engage the "-" polarity sign by shorting the *N ON* points; but since polarity is not necessary for our application, I left it alone. If you want a polarity display, solder a jumper between these two points.

With the modifications done, replace the board in the case and be careful that the LCD panel and rubber conductors stay in place. Return the five screws to their respective positions. The meter is ready for use as a 200-millivolt system.

Following the instruction for the CA-1 kit, solder all parts in place. Leave off the two DPDT pushbutton switches. Slide switches will replace these switches, with S1 as a DPST and S2 a DPDT. Also, there is an SPDT, S3, which chooses the input access.

For S1, solder each pole to the positive (red) lead of one of the two 9-volt battery connectors. One throw goes to the meter's positive lead, and the other goes to the CA-1's positive input.

S2 has the poles going to the center positions on the CA-1 board. The *high* throws connect to potentiometer R6 (see the instruction sheet).

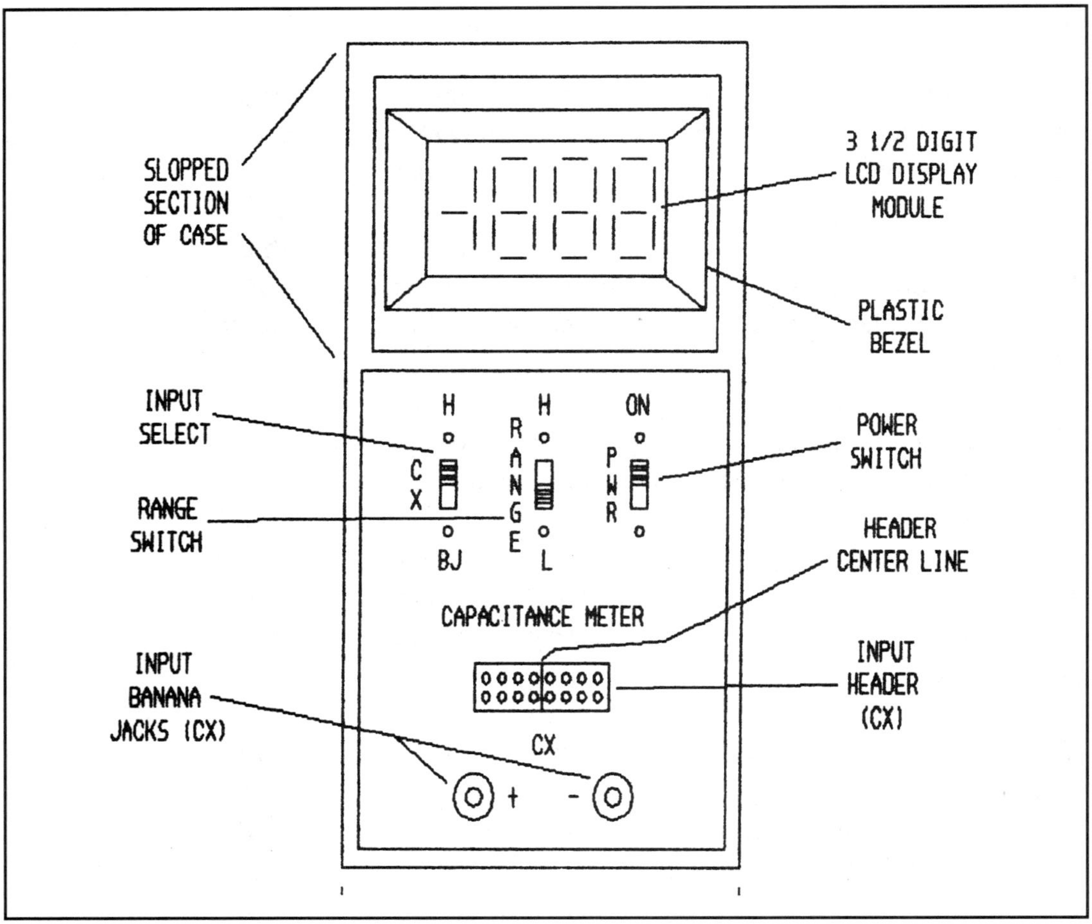

Figure 7-4. Capacitance meter panel template.

Only one of the *low* throws is employed, connected on the side of the board closest to diode D3.

The header jack is divided in half by soldering all pins on each half together. The soldered pins are then connected to a lead wire. This allows capacitors of varying lead lengths and spacing to be plugged into the device. You could employ the *speaker* connector that comes with the kit, but it would make it difficult to check caps with very short leads.

Wire together the negative (ground) side of the header and the black banana jack, and connect them to the negative (-) CA-1 input (J2). The red banana jack should go to one throw of S3, while the positive half of the header connects to the other. S3's pole is fed to the positive (+) input (J1) on the board. However, both the header and the banana jacks have to be mounted to the front panel before they can be soldered, so hold off on permanent connections for the moment. Actually, neither is really needed for a circuitry test.

Last, construct the voltage divider by soldering one end of each resistor (R10 and R11) together. Feed this union to the positive meter input. Wire the loose end of the 100-kilohm resistor to both the negative output (PL2) of the CA-1 and the negative input of the meter. The 1-megohm unused lead goes to the CA-1 positive output (PL1).

Now let's turn our attention to the case (*Figure 7-4*). I employed a generic handheld meter case, available from many suppliers and occasionally from the surplus market. This enclosure has flat front and a top that slopes toward the back end. Measurements are 5-3/4" (121 mm.) by 3-1/2" (88 mm.) by 1-1/8" (28 mm.) in the front, and 1-7/8" (48 mm.) at the rear. This gives the case ample room for both circuits and batteries.

Drill or cut the necessary openings for the switches, banana jacks and header. A sufficient aperture should already exist in the sloping area for the voltmeter module. If it isn't there, you will have to cut it.

The case will be sandwiched between the meter and an included bezel, so you may have to fashion a facing to fill in the empty space. That's what I had to do with my case. The facing can be made from sheet plastic and spray painted black to match the bezel. It can also be spray-painted red or whatever color you want. See, I'm easy to get along with!

Anyway, once you have made all the holes, spray-paint the flat surface with a light color (I used yellow for the prototype) so the black dry-transfer lettering will show up. You know what's next, right? Label the flat surface.

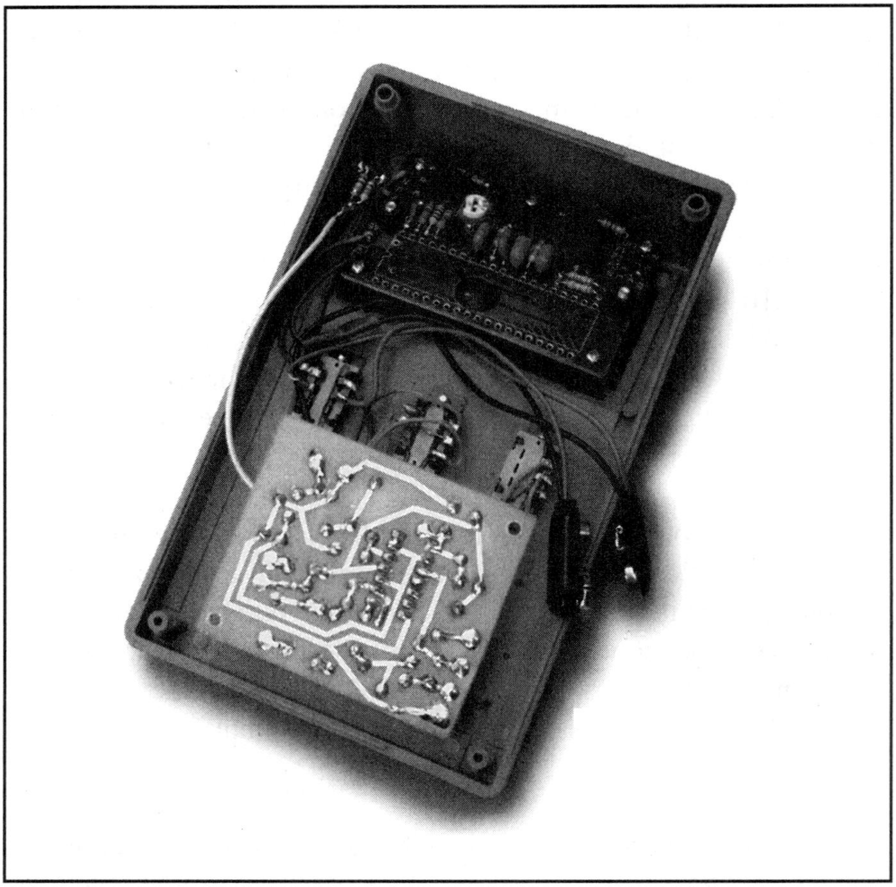

Photo 7-1. Internal shot of the completed capacitance meter.

I used *CX* for the input select switch (S3), with *H* for the header and *BJ* for the banana jacks. Other labels include *Range* for switch S2, *H* for high, and *L* for low. S1 is designated *PWR* for power, and *ON* is used for (you guessed it) on.

I used larger lettering for *capacitance meter* and *CX*, and used "+" and "-" to designate the test capacitor inputs. When you're done labelling your case, spray a light coat of clear enamel over the labelled area. Dry-transfer lettering will scratch off very easily if not protected.

Once the clear coat is dry, mount the meter, banana jacks, header and slide switches. Hook up the jacks and header. The unit is ready for testing and calibration (see *Photo 7-1*). Don't close up the case just yet, though.

Testing and Calibration

Testing the meter should be quick and clean. First install the batteries, then set range switch S2 to the low (L) position. Now, turn on the power (S1), and you should see a display of somewhere near *.000*. If there is no display at all, check to make sure the voltmeter is getting power. If the display is erratic, you might measure the output of regulator U1. It should be right at 5 volts.

As for calibration, with the *CX* inputs open, adjust potentiometer R6 until *.000* comes up on the LCD. This is a delicate adjustment, so it may take you a little time to hit zero. Actually, *.001* will be close enough.

Next, install a 100-picofarad capacitor (preferably mica) in the header and rotate potentiometer R1 (low setting) until a *.100* reading appears. Be sure that S3 is in the header *H* position. Last, set switch S2 (range) to high (H), and introduce a 1-microfarad capacitor to the header.

Adjust potentiometer R3 (high setting) for a *1.000* display. You should check several capacitors of the same type and value, using them to fine tune the adjustments to an average for all components. Also, test a double handful of marked (known value) caps to see if the meter is giving you readings close to what they should be. Don't panic if the

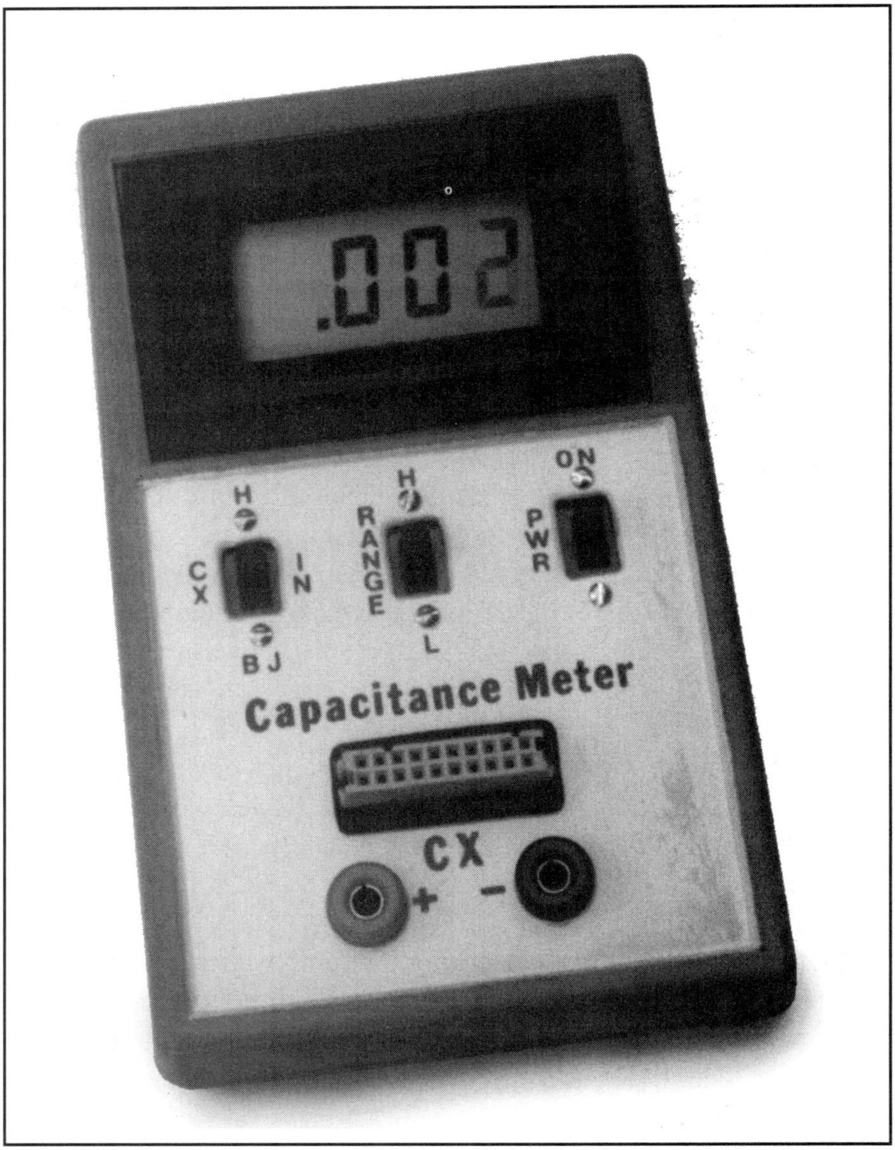

Photo 7-2. The completed capacitance meter in its case.

displayed values are a little off in one direction or the other, as most capacitors have a tolerance of about twenty percent.

You can also try a 0.001-microfarad unit in the low setting. This should give you a reading of *1.000*, or 1,000 picofarads. The same cap in a high value will display *.001*. Try several different values, like 0.0015 microfarads, 0.005 microfarads and so forth, to confirm proper meter operation. A 0.005 value will overflow in the low (L) setting.

If all goes well, your device is calibrated and ready to be put to work. Secure the CA-1 board to the bottom half (I used Velcro) and close up the case (see *Photo 7-2*).

Applications

As with most of the projects in this book, by the time we get to the *Applications* section, the device application is pretty obvious. However, let's talk a little bit about what this meter can do for you.

As stated previously, the digital capacitance meter will check a capacitor to see if it is functional (*Photo 7-3*). If the meter section is working and calibrated, and you see either no reading (display change) or an out-of-range (overflow) display, then you have a bad component. With an open capacitor, there will be no display change, while shorted caps will blow out the meter (*1.—*). Both situations are, of course, unacceptable.

So much for bad capacitors. Let's say you have a bunch of good units you want to test. For unknowns, set the meter range to whichever category you think the capacitor falls into (you know, high or low). Size can be helpful, as large-sized caps usually have larger values. However, this approach work only if you are familiar with the particular type of unit. A 0.1-microfarad monolithic is going to be about 1/10 the size of a 0.1-microfarad mylar. So, read up on the various unit types and their characteristics, especially appearance. If you don't get a reading in one setting, try the other. A 10-pico-

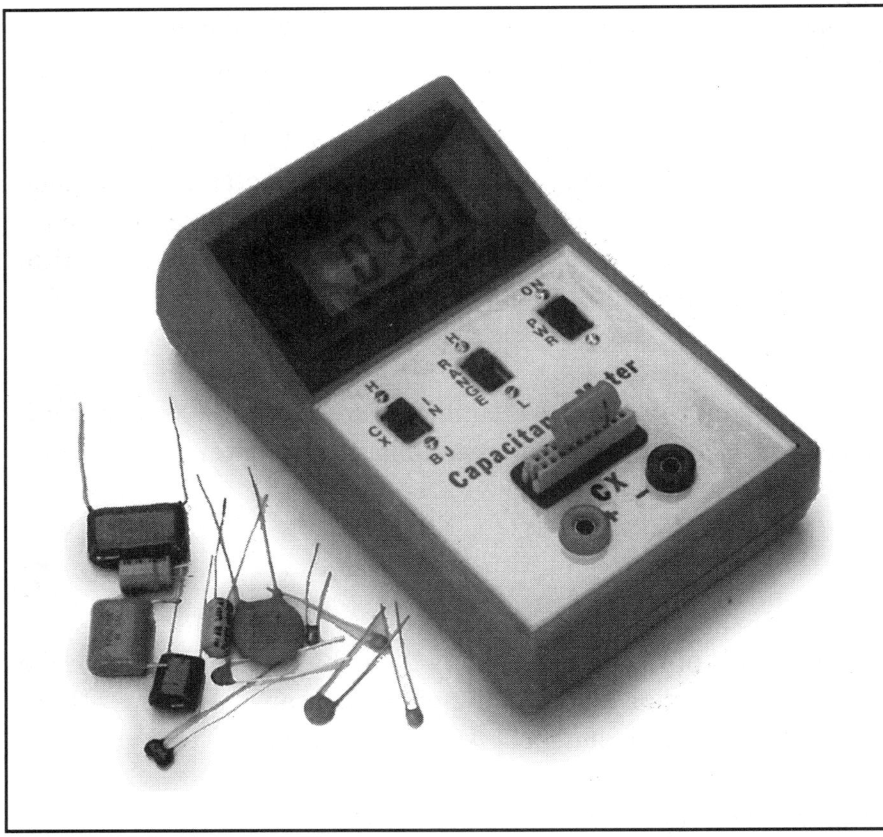

Photo 7-3. The completed capacitance meter in use, checking capacitors.

farad disk is not going to budge the meter if it is set for *high*. By the same logic, a 1-microfarad unit will overflow in the *low* setting. In short, don't give up on the little suckers until you have tried both ranges.

As for designated capacitors, the meter will help you confirm the actual value and confirm the unit's designation. This trait comes in very handy when you need closely matched components. It allows you to start with a pile of one value and ascertain which capacitors are best suited for the application.

Additionally, the meter can be useful for diagnostic purposes. Periodic tests of high stress capacitors can reveal flaws in circuit design by illustrating premature degradation of the component. This aspect is especially valuable when you are dealing with power supplies and computer equipment.

If you are like me, nothing electronic ever goes into the circular file until it has been cannibalized for parts. This is almost a religion on my part. Without a capacitance meter, though, I was never sure of a capacitors' condition. Most of the time, you can't really tell the shape of a capacitor by appearance. It is nice to know whether or not the caps going in the parts box do work.

On one last application note, when you use the banana jacks with cables, remember that the cables themselves possess a certain amount of capacitance. This won't show up in the high range, but when using the meter's low setting, plug in the cables and let the meter settle on a reading before trying to measure a small cap. Later, you can subtract the *cable only* reading from the final display.

Conclusion

Trust me! You're gonna like this meter. As a matter of fact, after using it for a while, you will learn to love it. This is one of those instruments you can get along without, but it is soooo much nicer to have it available. Not only will this capacitance meter save you a bundle on surplus capacitors, you will always know the exact values you are dealing with. When it comes to prototyping, this can be a big saver of the green stuff.

In another vein, there are few frustrations more intense than struggling with a temperamental high-frequency RF circuit only to find that a ceramic disk marked *30 picofarads* is really 10 picofarads. No wonder the thing was off frequency! Makes you want to get real angry with someone. Anyway, with this capacitance meter in hand, this problem should never plague you again. Besides, the meter is fun and easy to build, and as the saying goes, "It'll keep you off the streets." Well, maybe.

Chapter 8
Universal Four-Digit Counter

Introduction

This device will be far more handy than you might imagine. Using its on-board clock, the unit can count in either seconds or minutes (or anywhere in between). The external input allows for the introduction of pulses from other sources. Plus, the count can be up or down.

Additionally, a four-digit binary-coded decimal (BCD) port provides both input and output capability. Here, in the input mode, thumbwheel-style BCD switches can preset a starting point and have the counter move on from there. As for binary output, the prototype utilizes this feature to light up small light-emitting diodes (LEDs). In this way, you have both digital and binary displays. Not only is the counter a useful stopwatch, it will serve you well when you're developing new preconditioning or gate circuits for frequency counters, or event (unit) counter projects. This system contains decade counters/dividers, decoders and drive circuits, all ready and waiting to be employed. This leaves you free to concentrate on your new circuit's design.

With that said, let's take a closer look at what makes this gem tick. Universal counters are not readily seen on the commercial market (at least not anymore), but they are a valuable addition to any workbench.

Theory

The secret behind the four-digit counter is another astounding Intersil/ Harris chip from the 1970s and '80s. This one, the ICM7217, puts all

UNIVERSAL COUNTER PARTS LIST

SEMICONDUCTORS

U1	ICM7217IJI 4 Digit CMOS Up/Down Counter (Jameco)
U2	7805T Positive Voltage Regulator
U3	74LS04 Hex Inverter
U4	74LS00 Quad 2-Input NAND Gate
U5	LM555 Oscillator/Timer
LED1	Yellow "Top Hat" Light Emitting Diode
LED2-5	Red, Yellow, Green T-1 Light Emitting Diodes
LED6	Green T-1 3/4 Jumbo Light Emitting Diode
D1	1N4733 5.1 Volt Zener Diode
BR1	Bridge Rectifier (WO4M, or Equivalent)
DSPL	4 Digit Common Anode LED Display (2-MAN6710, Etc.)

RESISTORS

R1	10,000 1/4 Ohm Watt Resistor
R2	1,000,000 Ohm Panel Potentiometer
R3	1,000 Ohm 1/4 Watt Resistor
R4	220 Ohm 1/4 Watt Resistor

CAPACITORS

C1	2,200 Microfarad Electrolytic Capacitor
C2	1 Microfarad Electrolytic Capacitor
C3	2.2 Microfarad Tantalum Capacitor
C4	100 Microfarad Electrolytic Capacitor
C5	0.01 Microfarad Mylar Capacitor

OTHER COMPONENTS

S1, 6	SPST Pushbutton Switches
S2, 3	SPDT Pushbutton Switches
S4, 7	SPDT Slide Switches
S5	Momentary Contact Pushbutton Switch (NO)
S8	SPST Slide Switch
T1	120 VAC to 12 VAC Power Transformer
J1, 2	Banana Jacks

TERM1 Dual 8-Position Pin Jack
P1 120 VAC Line Cord and Plug Combination

MISCELLANEOUS

PCB Materials, Solder, Hookup Wire, Hardware, Case (Instrument Style), Paint, Dry Transfer Lettering, Knob, Etc.

of the goodies you need into a standard 28-pin DIP package that runs on a single, positive 5-volt supply.

7217s come in four varieties, each distinguished by the type of display addressed (common anode or cathode). The ICM7217IJI, used in this prototype, supports common-anode LEDs and a decade display of *9999* maximum. The AIPI uses common-cathode elements with a decade readout. The BIJI and CIPI both provide timer displays of *5959* maximum (the first addressing common anode and the later common-cathode arrays).

Actually, there is another set of four chips; the 7227 series. These chips are basically identical in performance to the 7217s, but are intended for microprocessor-based systems. They are also rather hard to find these days.

Back to the 7217s. In each case, the chip furnishes everything needed to count the incoming pulses, decode the counts, and display them on LED readouts up to and including instrument size (0.56"). The output is of the multiplexed variety; only eleven lines are required for the seven segments and four digits.

The multiplexing rate is 2.5 kilohertz when free-running, but it can be lowered through the addition of a single external capacitor. This is accomplished by *scan* pin 13, with the capacitor moving from there to the positive line.

As mentioned, a four-digit BCD port (pins 4-7) allows binary information to come and go. This works in conjunction with pins 11 and 12; the *load resistor* (LR) and *load counter* (LC), respectively. For example, when both LC and LR are open, the port is in the output mode; however, if either is taken high, the port becomes an input. Furthermore, if LC is high, the count is stopped and transferred to the counter. If LR is high, the count is multiplexed into the register but the counter continues. When both are high, the count is halted, and the register and counter are loaded with the last count. So, the versatility of the BCD lines is more than apparent. All are accessible from a front panel header and are quite useful during certain projects.

Another nice feature of the 7217 is the *carry/borrow* pin 1. This allows for the automatic cascading of additional counter ICs; thus, you can easily have an 8-, 12- or even 16-digit counter. Whenever the count reaches 9999 or 5959, the carry/borrow issues a positive pulse that initiates the next chip.

Pin 2, *zero*, becomes negative if the count is 0000. The prototype counter indicates this condition with a panel-mounted LED. The last function of this nature involves pin 3, the *equal* line. If the counter and register have the same value, the pin goes low. Here again, a panel LED could be installed to note the occurrence.

Control of the display is handled by pin 23 in a three-state arrangement. When pin 23 is low, a full display is present. If pin 23 is floated, leading zeros will blank out; and if taken high, the entire display is inhibited. This last scenario does allow the display to be used for other purposes, but the prototype shows only the first two conditions.

The incoming pulses are sent to pin 8, the *count input*. The counter is incremented, or decremented, by a rising edge signal. Pin 9 is the *store* which controls the latches. When brought low, the store transfers the counter's contents to the latches and ultimately to the LED display. The *reset* function is handled by pin 14, and as might be expected, the pin resets the counter to *0000* when brought low. In the prototype, all of these functions are managed by a two-chip logic array, but more will be said about that later.

The final control is pin 10. It determines the direction of the count. For example, when high or floating, the 7217 counts up; when brought low, it counts down. Another handy feature! For the record, this is called the *up/down* line. Gee, who would ever have thought of that?

Anyway, that's the ICM7217 in a nutshell. But what a nutshell! (Check out *Figure 8-2* for the pinout). However, nothing's perfect, and with

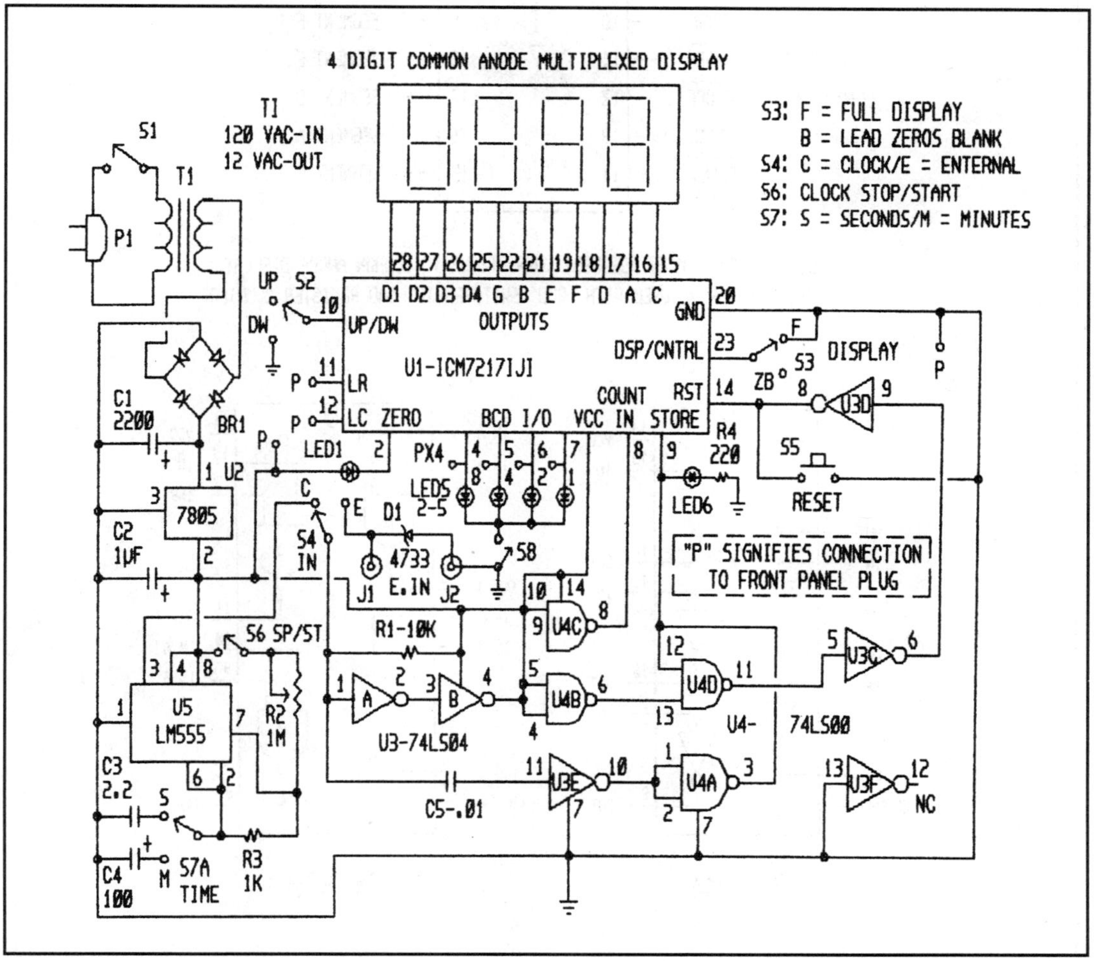

Figure 8-1. Universal four-digit counter.

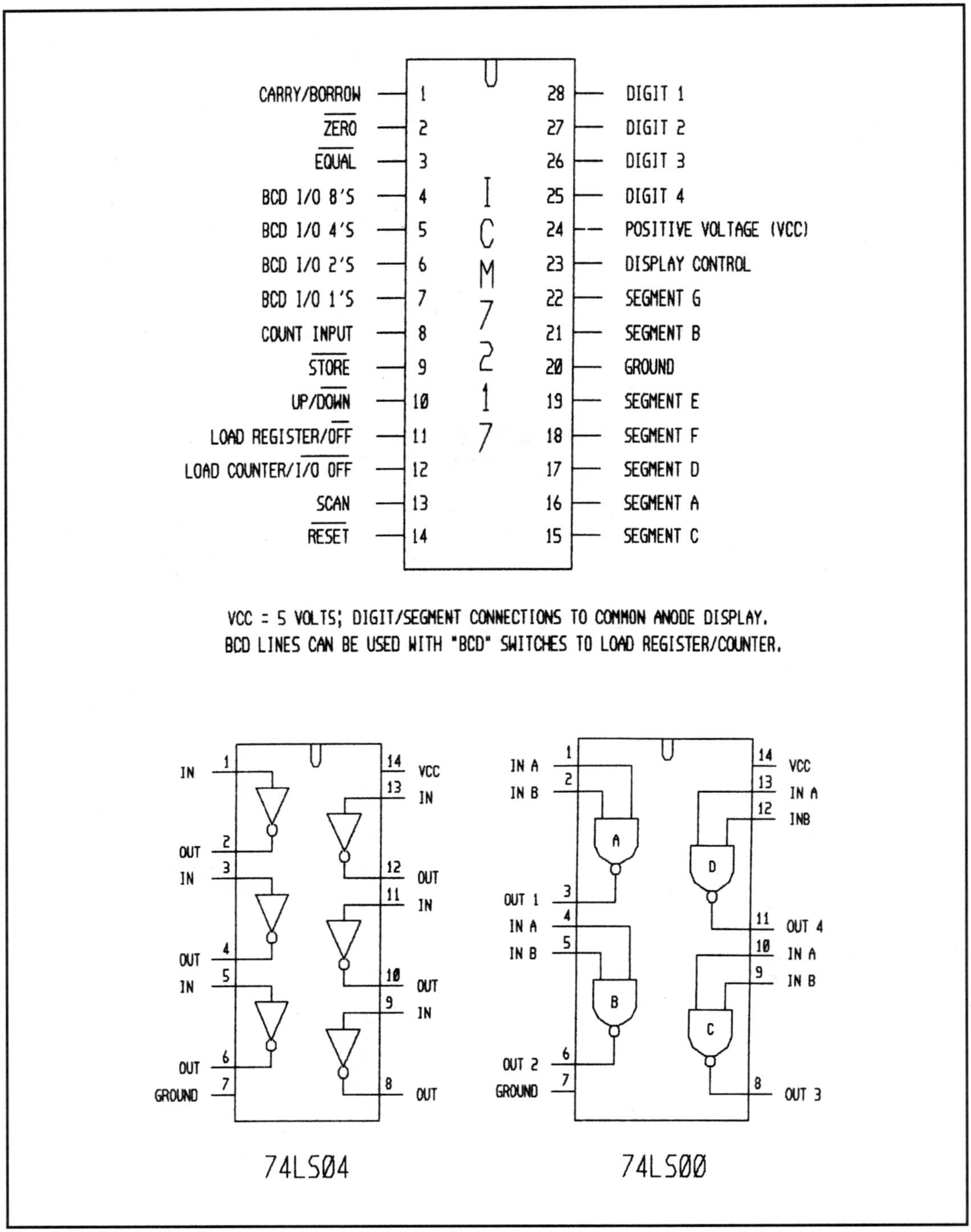

Figure 8-2. ICM72171J1 pinout, and the 74LS04 and 74LS00 pinouts.

the 7217, frequency capability is a slight drawback. The chip is guaranteed for up to 2 megahertz, and will usually run to 5 or 6 megahertz without difficulty. So, in today's high-frequency circuits, some prescaling will probably be necessary in order for the 7217 to handle most frequency counting operations.

Our next area of concern is logic control. Referring to *Figure 8-1*, you can see that logic control is accomplished with inverters and NAND gates. Pretty basic in nature. Rather than go through each gate and specifically describe what each one does, let me just say that the logic is designed to produce the proper signal level at the reset, store and count input lines.

If you enjoy tracing logic circuits, or just have an urge to do so, employing the truth tables will help you confirm this section's operation. At the end of the trail, during counting, you will want to see highs at all three pins and lows when the chip completes a count. The low levels store the count in the latches and reset the IC, preparing it to evaluate the next input signal.

As for the chips themselves, I went with the transistor-transistor logic (TTL) 74LS00 (U4) and 74LS04 (U3). Complementary metal oxide semiconductor (CMOS) ICs could be used, but at 5 volts, they don't run well above 2 megahertz. Also, since you may want to engage the counter as the basis for a frequency meter, this is not a good deal. The TTLs will go as high as 20 megahertz; thus, for me at least, they are the better choice. Momentary contact switch S5 will manually reset the counter, and switch S2 will allow for either an up or down count. These functions can be implemented at any time during normal operation, as can the display control, switch S3.

S4 selects the input. It offers a choice between the on-board LM555 (U5) astable clock circuit, or an external source by way of a pair of panel-mounted banana jacks (J1 and 2).

The clock is a standard 555 configuration, with switch S7 determining the general oscillation scope. This is accomplished by the size of the timing capacitor (C3 or C4). The larger the capacitor, the slower the rate. The two values chosen provide a *seconds* range with C3 and *minutes* with C4.

Potentiometer R2 adjusts the oscillation to exactly what you require within the confines of the capacitors, and switch S6 acts as a *start/stop* control by disconnecting R2 from the positive voltage. For the prototype, I found C3 provides everything from hundredths of a second to about 5-1/2 hours, while C4 furnishes around 1/2 second to 1-3/4 years. So far, I've yet to find a need outside those parameters.

The display can be any four-digit multiplexed common anode array. I had such a display on hand from an old surplus purchase (a Litronix 4530). But the display could be two dual-digit displays, such as MAN6710s wired in multiplex, or four individual LED units.

Multiplex merely means connecting all *a* segments, *b* segments, *c* segments, and so forth, together and running a line to the appropriate 7217 pin. Each of the four common anodes are wired to the 7217 digit lines (D1-D4), with D1 being the least significant digit (LSD) and D4 the most significant digit (MSD).

I always leave the poor ol' power supply for last, but not because I have less respect for it than for the rest of the circuit. There is little, if any, substitute for a good power source. In this case, as with all test equipment, we want a regulated system that not only provides a steady voltage but also enough current. The 7217 will need around 200 milliamps, depending on the display condition (the number of segments lit). The rest of the system will probably use somewhere in the neighborhood of 100 to 150 milliamps. This means you will want to supply a minimum of 500 milliamps.

Regulator U2, a 7805T, will furnish up to 1.5 amps. In the prototype, I also used a 12-volt transformer that supplies about 1 amp. Between the two, you should have all of the power you need, and then some.

Employ the old standby linear configuration, with bridge rectifier BR1 and capacitor C1 handling rectification and filtering. Capacitor C2 provides additional filtering and transient suppression, and prevents the 7805 from oscillating. LED6 and resistor R4 form the *on* indicator network, while S1 is the *power on* switch.

You should observe the usual precautions, such using an adequate heat sink for U1 and wire nuts for all 120 VAC connections. Also, you can incorporate a 1-amp fuse in the transformer's primary winding circuit, if desired. Simply connect it in series with one side of the primary.

That about covers the operation of the 7217 counter. Not too hard, is it? Now let's see how the prototype went together.

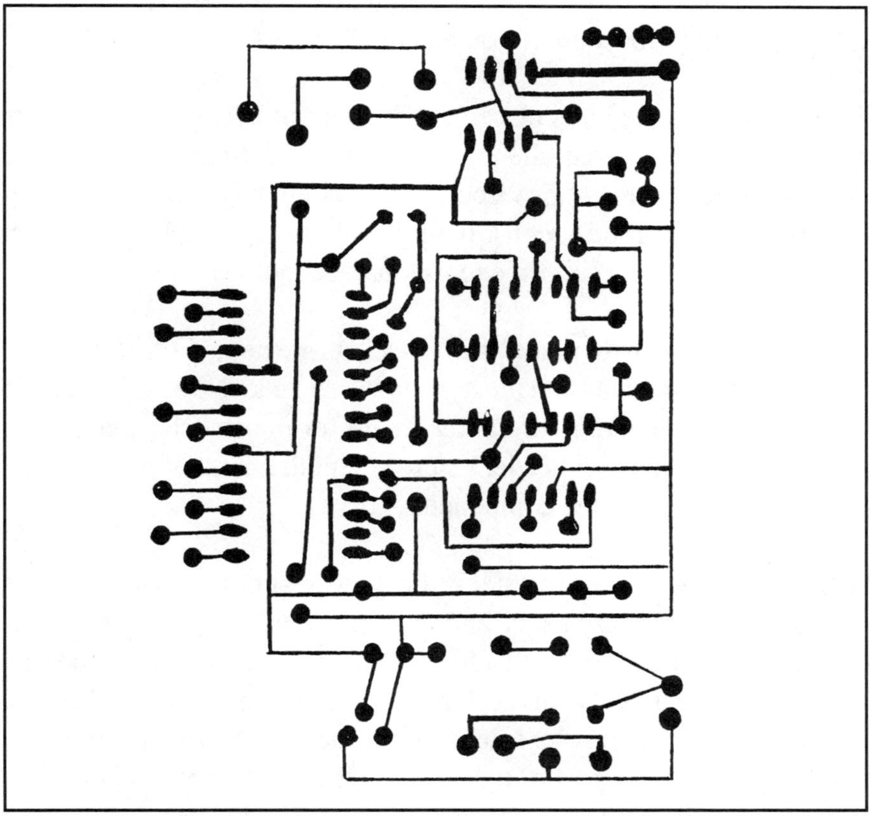

Figure 8-3. Universal counter foil pattern.

Construction

Figure 8-3 illustrates the foil pattern for a printed circuit board (PCB) suitable for this project. I do recommend using a PCB, but I almost always recommend PCBs. I find them easier to work with and more reliable. However, in the past I have received some arguments against this, as some individuals prefer point-to-point wiring on perfboards. Well, what can I say? There is no accounting for some people's taste. I mean, really! I'm just kidding. This project will work fine using point-to-point wiring.

For the prototype, I employed a 4-3/8" (102 mm.) by 3-1/8" (82 mm.) printed circuit board. This size accommodates all components and leaves room for mounting holes that match up with the cabinet standoffs. Anyway, we will examine this approach, but basically there is little difference in terms of procedure if you elect to use perf board.

First, referring to the parts placement diagram in *Figure 8-4*, install the jumpers. Note that two of the jumpers (J10 and J16) go under IC sockets. Due largely to the logic circuits, this system has quite a few jumpers. So, if you are good at constructing double-sided PCBs (which I am not), this would be a good time to use that technology.

Once the jumpers are in place, the IC sockets are next. Follow this with the discrete components and the regulator/heat sink. Now, solder appropriate lengths of hookup wire to the board for the switches, jacks, display, LEDs and junction header. Here again, the number is substantial due to the panel-mounted components.

At this point, it's wise to check the power supply. Hook up the transformer's secondary leads, and test regulator U2, pin 2, for 5 volts. Next, the IC sockets should display the positive 5 volts at the following positions: U1, pin 24; U3 and U4, pin 14; and U5, pin 4/8. If all is well, connect the panel components, except for the power switch and banana jacks. They have to go through the panel first, so hold off on them. Now, plug in the integrated circuits, and be sure to observe proper orientation.

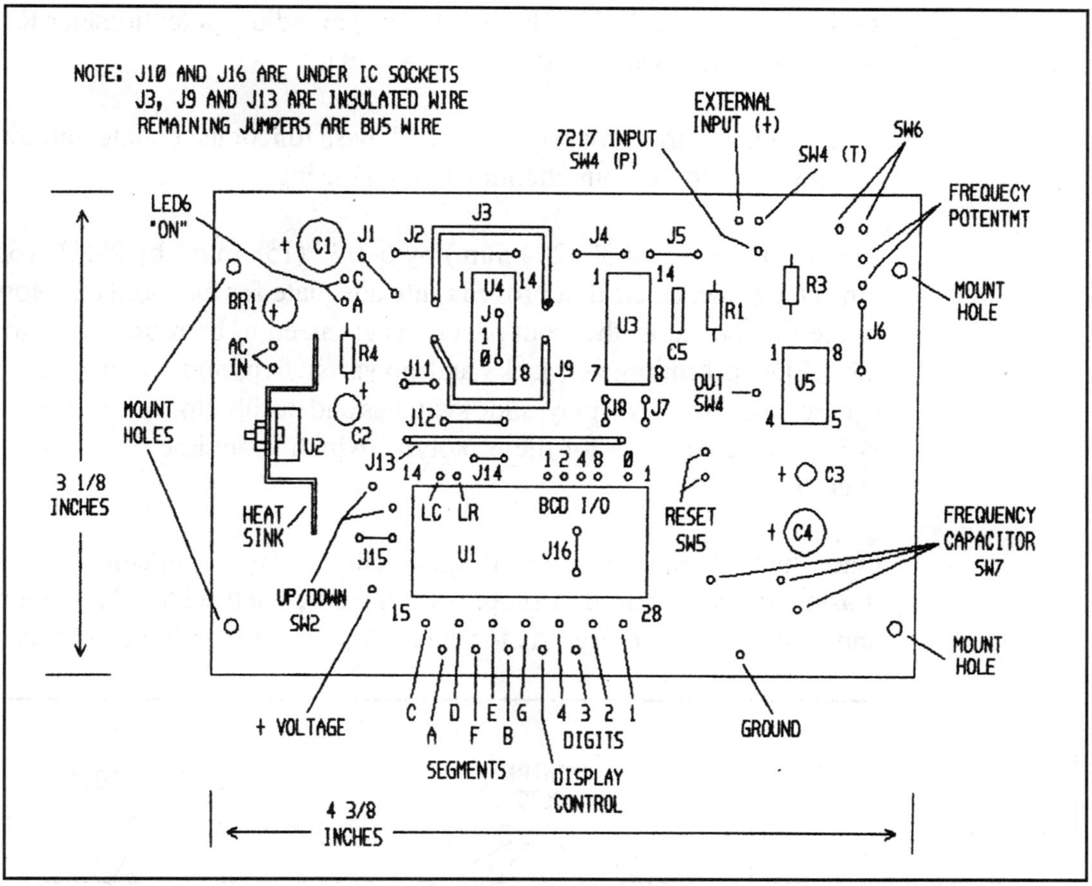

Figure 8-4. Universal counter parts placement.

With the cord plugged into the AC wall jack, and the leads for the power switch temporarily shorted, set the *stop/start* switch (S6) to the *stop* position, and hit the power switch. The power *on* LED should light, and the display will show *0000*.

Occasionally, the display will read *2222* or *3333*, but this seems to be normal since it will change to *0000* on reset. Also, the *zero* LED should illuminate because the counter is at the zero state, or at least it thinks it is.

Next, with clock switch S7 in the *seconds* position, change S6 to *start*. The display will begin to count either up or down, depending on the setting of switch S2. You'll note that as you adjust potentiometer R2, the counting rate will speed up or slow down.

If everything works properly, you now have a functional counter circuit. It's now time to turn our attention to the housing.

I selected a standard 8" (204 mm.) by 6-1/4" (158 mm.) by 2-1/2" (63 mm.) instrument case that proved quite adequate for this project. Most of the fuss concerns the front panel. There need to be round holes for the LEDs, potentiometer, jacks and power switch, and rectangular or square holes for the display, slide switches and pushbutton array. *Figure 8-5* shows a template for the prototype, which should help you locate everything.

Once the front panel is prepared, spray paint it with a light color and set it aside to dry. Then, if desired, paint the case shell with a dark color, and again allow it to dry. As for the back panel, it merely needs a hole

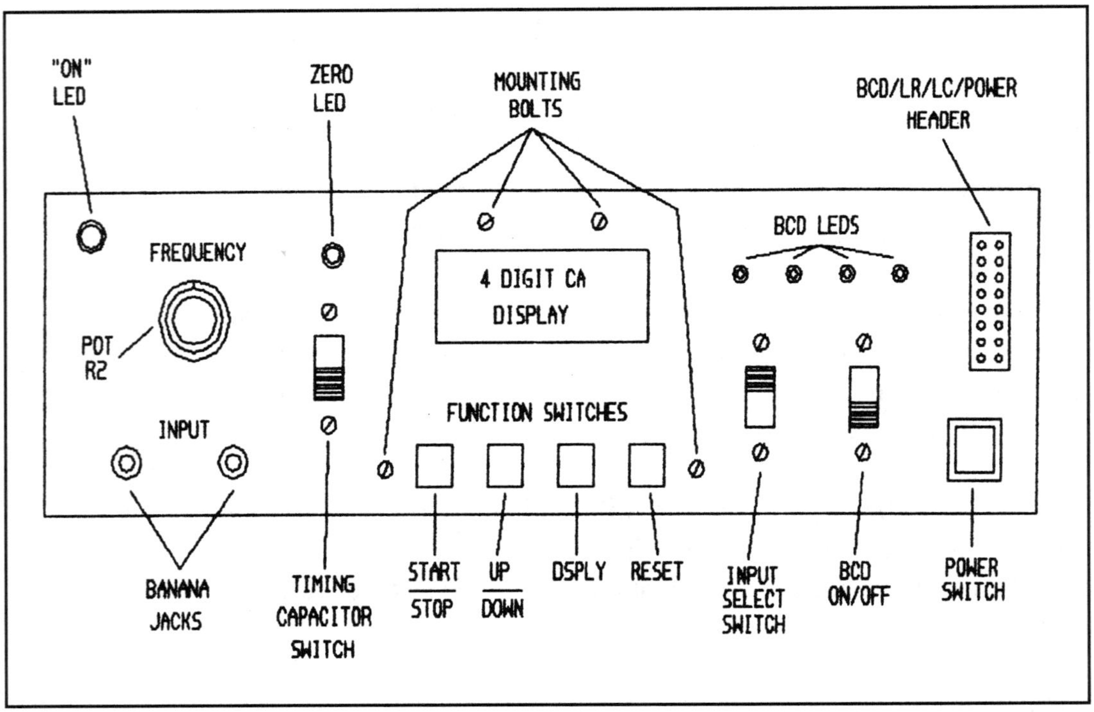

Figure 8-5. 7217 counter front panel template.

large enough to accommodate a rubber grommet for the power cord. Painting this panel really isn't necessary.

For our next trick, the cured front panel needs to be labelled with dry-transfer lettering, then sprayed with a light coat of clear enamel. The clear coat protects the delicate labels.

Take a breather. The next step is to install all of the panel parts and wire them to the hookup leads. Very little needs to be said about this procedure other than you should approach it carefully. I used multi-colored or rainbow ribbon cable for the display to help identify the individual segments and digits. While this is convenient, it's pure laziness on my part. If you're cautious, any type or color of wire will work.

The access header, zero and BCD LEDs are cemented to the panel with epoxy, while everything else has its own required method of installation. For example, the display and pushbutton array are secured with bolts through existing mounting holes in each.

Once you have everything in place and connected, you're done (*Photo 8-1*). Let's try out the thing.

Testing and Operation

Regarding testing, we've pretty much done that already. It is, however, advisable to run through the controls a second time to confirm proper operation. I wish I had a dime for every time a project worked outside the case but rebelled when placed inside. It supposedly has something to do with all those gremlins lurking about. An as-yet unconfirmed explanation, though.

If all is well, you are ready to put the counter to work. At this point, you probably already have a good idea of how to do that. The testing procedure demonstrates most of the functions, but let me review what's been covered, and hit a few of the more subtle points.

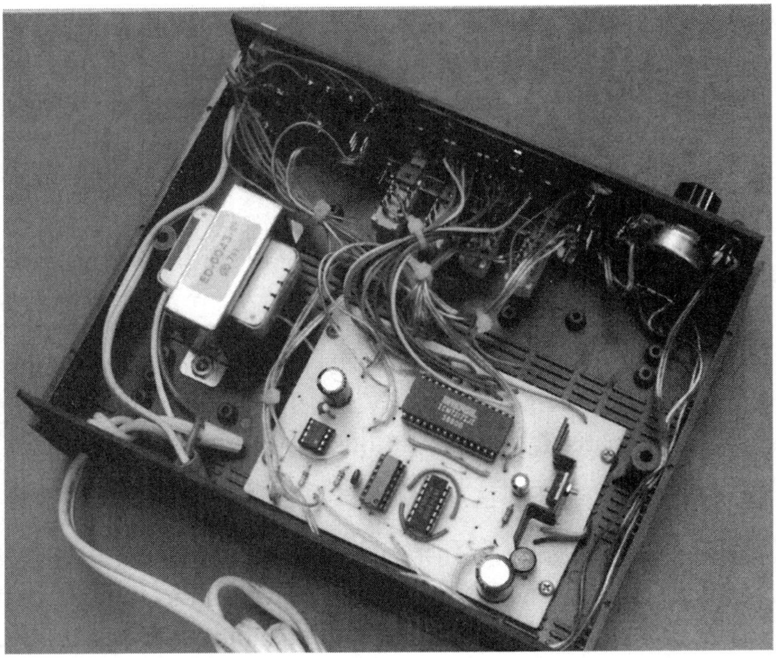

Photo 8-1. Interior view of the universal BCD counter.

Starting with the four control switches, S6 stops and starts the internal 555 clock. S2 determines the direction of the count, either up or down. S3 allows for full display or leading zero blanking. Last, S5 manually resets the counter to *0000*.

When the 555 clock is employed, potentiometer R2 (marked *freq*) fine tunes the timing period established by capacitor C3 or C4. Switch S7 selects which capacitor is engaged. This permits you to obtain the exact count you need.

Concerning the input to the 7217, this is handled by switch S4. Either the clock or an external pulse can be introduced to the circuit, with the external signal coming in through banana jacks J1 and J2. The latter allows the four-digit counter to function as the basis for a variety of pragmatic designs.

The last section involves the binary-coded decimal (BCD) port. As described earlier, this port can act as an input or an output. In each case, the versatility of the 7217 (hence its application) is vastly expanded. Think about it! The ability to send and receive binary codes gives rise to a variety of possibilities.

The prototype is set up in the output mode. It utilizes four T-1 LEDs to indicate the code for each digit that appears on the display. You'll note, after some observation, that the BCD display exhibits the first or least significant digit (LSD) until the display changes to a higher digit. It shows the value of the higher digit for one count, then returns to the LSD.

Is that confusing? It kind of confuses me. Let's look at an example. If the display is *0007*, then the *4*, *2* and *1* BCD LEDs will be lit. If the display is *0005*, then *4* and *1* will light up, and so forth. However, when the count reaches *10*, only the *2* BCD LED will illuminate.

Are you with me so far? Good! On the *11* count, only the *1* LED will light up; then the BCD sequence will resume single digits until it reaches *20*. On that count, just the *2* will glow. *2* and *1* will light up for *30*, *4* will light up for *40*, and so forth.

Did that help? I hope so, because I think I'm more confused than ever. Just play with the counter for a while and you'll catch on. Honest, it isn't as complicated as I have tried to make it.

Incidentally, switch S8 disconnects the BCD display. The header on the front panel also provides access to the BCD port and the control lines. This allows you to introduce or receive information from the port through the manipulation of the load counter (LC) and load register (LR) lines. By jumping either or both to the positive terminal, you can control the actions of the BCD port. Those actions include input, output, counter and register status, all of which will be useful when you're prototyping new designs. Of course, the four BCD lines are also accessible from the header.

So, when working with the 7217, it's prudent to recall that any BCD-producing device can control the counter, and any device regulated by binary codes can, subsequently, be managed by the 7217. For example, the BCD outputs could be used in conjunction with, say, the TTL 74154 4- to 16-line decoder/demultiplexer. In this arrangement, each count would activate one of the 16 74154 outputs, which could be used to control a multitude of other devices, circuits or systems. Think in terms of this approach, and I'm sure several other similar applications will come to mind. All in all, the existence of the BCD port can be far more practical than it might at first appear.

Conclusion

That, my friends, is the saga of the 7217 universal counter (*Photo 8-2*). As I said at the outset, benchtop counters are no longer a common commercial item, which is truly a pity. As this chapter has hopefully illustrated, universal counters do have a legitimate place in your test lab. Additionally, the knowledge learned here will provide you with a valuable tool for designing your own 7217 system. Any time you need an up/down counter that doesn't need to operate above 5 megahertz, the ICM7217 is your boy. Features like chip cascading and binary control

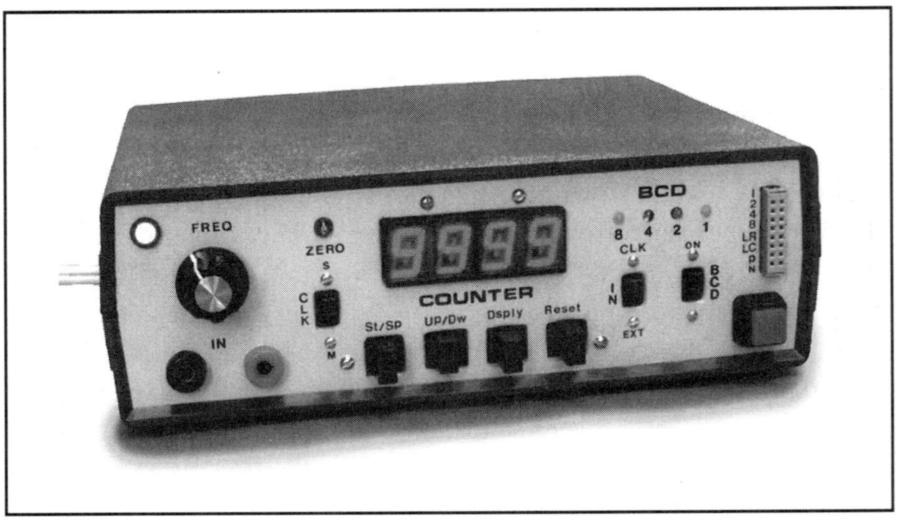

Photo 8-2. External view of the completed universal counter.

put this IC in a class all its own, while the price (in light of performance) is anything but oppressive. Try designing a similar system using individual components, and I think you will quickly appreciate the 7217's price tag.

On a final note, I especially enjoy this project because of a fascination with counters. They aren't particularly complex devices, but few of today's electronics could function without them. Besides, those constantly-changing pretty, bright numbers always dress up a prototype.

Chapter 9
The Workbench Digital Multimeter

Introduction

While most of us use a handheld digital multimeter in our electronic exploits, this bench version of the device comes in very handy. For one thing, it sits firmly on the bench or a shelf and doesn't getting in the way. I don't know about you, but I'm forever having to reposition my handheld unit while I'm working. Portable meters also slide around a lot, usually when you run out of lead length. A couple of times, mine has even ended up on the floor. Not the best of situations.

Don't get me wrong; I'm not knocking the small devices. I mean, what would we do without them, right? But for many workbench projects, the unit we are about to build is far more practical. As for operation, it is virtually identical to its portable cousins, the only differences being a light-emitting diode (LED) display and an AC-based power supply. For me, at least, the LEDs are easier to read, and powering the unit from the wall eliminates possible erroneous readings due to low batteries. So, let's take a look at how this multimeter works and how to put it together.

Theory

As I promised in an earlier chapter, we will again be working with the ICL7107 analog-to-digital (A/D) converter chip. Personally, I can't say enough nice things about this integrated circuit (IC), although I do run into some arguments from time to time.

BENCH MULTIMETER PARTS LIST

SEMICONDUCTORS

U1	ICL7107 3 1/2 Digit A/D Converter
U2	7805T Positive 5 Volt Regulator
U3	7905T Negative 5 Volt Regulator
LED1	Green T-1 3/4 Jumbo Light Emitting Diode
DSP1	3 1/2 Digit Common Anode LED Display (See Text)
BR1	Bridge Rectifier (W04M, or Equivalent)
D1, 2	1N4007 Rectifier Diodes
D3	3.9 Volt Zener Diode (1N4730, 1N746, Etc.)
D4	5.1 Volt Zener Diode (1N4733, 1N751, Etc.)
D5-8	1N914 Signal Diodes, or Equivalent

RESISTORS

R1	1,000 Ohm Multi-Turn PCB Potentiometer
R2	22,000 Ohm 1/4 Watt Resistor
R3, 5	100,000 Ohm 1/4 Watt Precision Resistors
R4, 12, 13, 15	1,000,000 Ohm 1/4 Watt Precision Resistors
R6, 10	10,000 Ohm 1/4 Watt Precision Resistors
R7, 9	1,000 Ohm 1/4 Watt Precision Resistors
R8	100 Ohm 1/4 Watt Precision Resistor
R11	223,300 Ohm 1/4 Watt Precision Resistor
R14	2,700,000 Ohm 1/4 Watt Resistor
R16	0.1 Ohm 5 Watt Power Resistor
R17	1 Ohm 5 Watt Power Resistor
R18	10 Ohm 10 Watt Power Resistor
R19	100 Ohm 10 Watt Power Resistor
R20	470 Ohm 1/4 Watt Resistor
R21	47,000 1/4 Watt Resistor
R22	220 Ohm 1/4 Watt Resistor

CAPACITORS

C1	0.22 Microfarad Mylar Capacitor
C2	0.47 Microfarad Mylar Capacitor
C3	0.01 Microfarad Mylar Capacitor
C4	0.1 Microfarad Mylar Capacitor

C5	100 Picofarad Mica or Ceramic Disk Capacitor
C6, 7	0.1 Microfarad/250 Volt Mylar Capacitors
C8, 10	330 Microfarad Electrolytic Capacitors
C9, 11	3.3 Microfarad Electrolytic Capacitors

OTHER COMPONENTS

T1	120 VAC to 12 VAC Power Transformer
S1, S2	8- 6PDT Pushbutton Switches (Alps # 1-211-11015C- All Electronics, Altronics, Mouser)
S3	1- 6PDT Pushbutton Switch (Only 1 Section Used)

MISCELLANEOUS

Solder, PCB Materials, Switch Caps, Hook-Up Wire, Hardware, Clear Red Plastic, 8" x 6 1/2" x 2 1/2" Instrument Case, Dry Transfer Lettering, Paint, 120 Volt Line Cord & Plug, IC Sockets (40-Pin), Heat Sinks, LED Holder, Etc.

As a quick review, the 7107 contains a dual-slope analog-to-digital converter, clock, reference, digit decoders and display drivers, all providing the basis for a simple yet accurate voltage measuring system. The prize in this box is a digital display of the results, courtesy of the decoders/drivers.

Once complete, the relatively simple 7107 circuit is capable of evaluating a direct current (DC) voltage and converting it into a nice, easy-to-read digital format. This, of course, provides a multitude of possibilities. More (a lot more) will be said about this down the line.

Since I have already gone through the operation of the dual-slope A/D converter, and I'm certain you have a thorough understanding of the process, I will not bore you with a review. If you are still unclear on A/D functions, review Chapter 2. Instead, let's tackle a second challenge of this project; namely, the rest of the circuitry. Yes, I know this sounds ridiculously simplistic, but remember that the 7107 can measure only voltage. Thus, to appraise resistance or amperage, we have

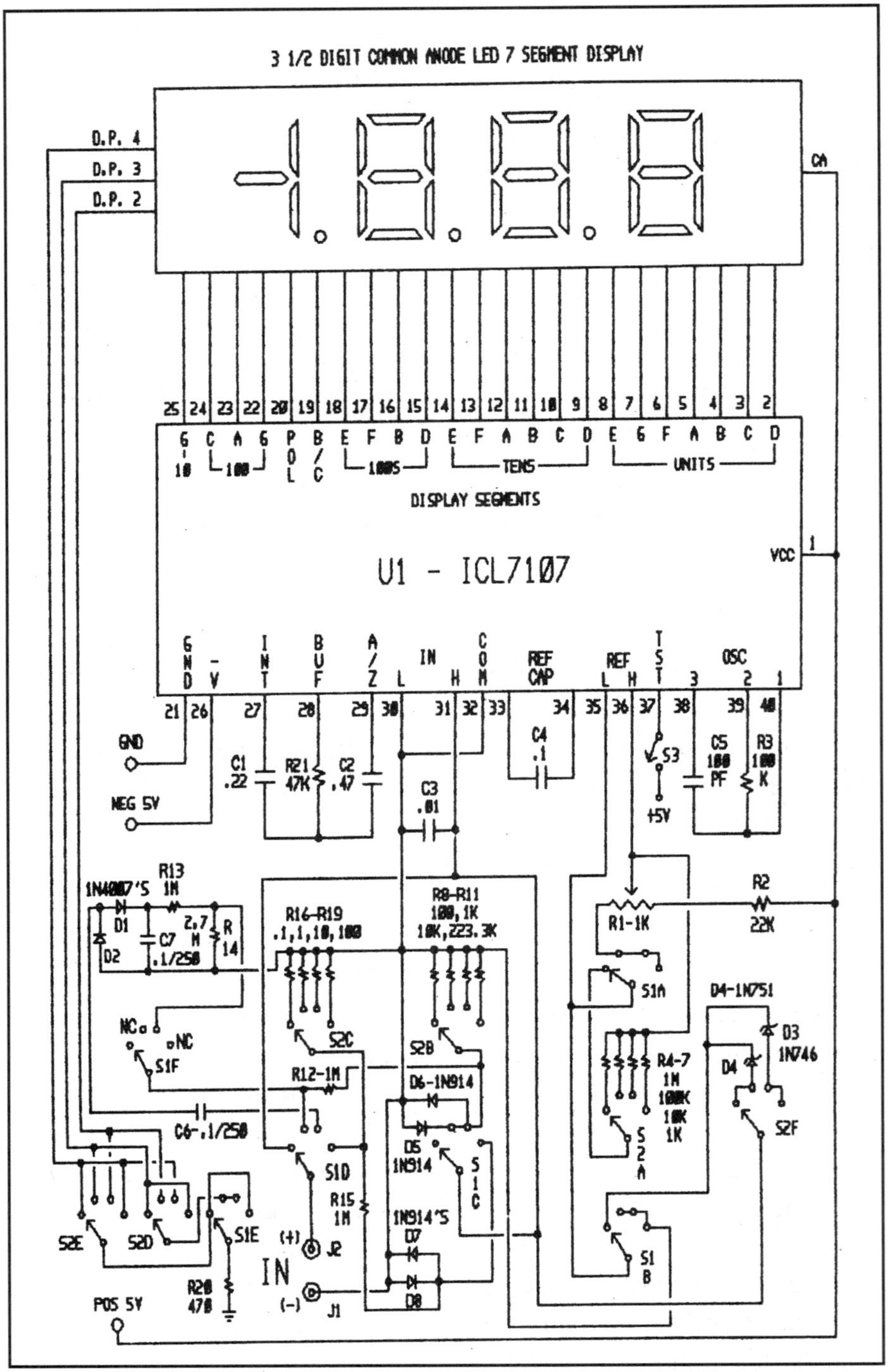

Figure 9-1. Bench multimeter schematic.

to effectively generate a voltage representative of the value of those commodities. Let me clarify this: When dealing with voltages, there is not a problem. The 7107 meter reads voltage, a maximum of 200 millivolts, and potentials that are easily handled with voltage dividers. Alternating current (AC) is first rectified then applied to the input.

Looking at the schematic in *Figure 9-1*, you can see that the voltage division comes from resistor R12 and a second resistor selected by switch S2B. Each *second* resistor (R8, R9, R10 and R11) is chosen to provide the proper division for the range settings of 2 kilovolts, 200 volts, 20 volts and 2 volts, respectively.

Consequently, the dividers take the input voltage and reduce it to the 0- to 200-millivolt range. It is then evaluated and displayed by the 7107. But what about ohm readings?

As stated, we have to develop a representative voltage that the A/D understands, and the easiest way to do this is to alter the 7107's reference voltage. The unknown resistance can then be sent to the chip's high and low inputs, as well as used in series with the newly-created reference. The purpose of this maneuvering is to generate the needed representative voltage across the meter's inputs. The 7107 does its magic and digitally displays the value of the unknown resistor. Neat, eh?

Again referring to *Figure 9-1*, switch S2A determines which of four resistors, R4, R5, R6 or R7, replaces potentiometer R1. This action sets the new reference voltage. Switch S2F selects D4, a 1N4733 zener diode, which provides a regulated 5.1-volt potential to the pole of S1A and the reference low. The regulated voltage helps maintain accuracy in the measurement. All switch connections are shown in *Figure 9-2*.

Note that S2F's pole is connected to both the positive input (J2) and the 7107's high input. This arrangement, in concert with J1 (ground), furnishes the *series* connection that completes the resistive measuring

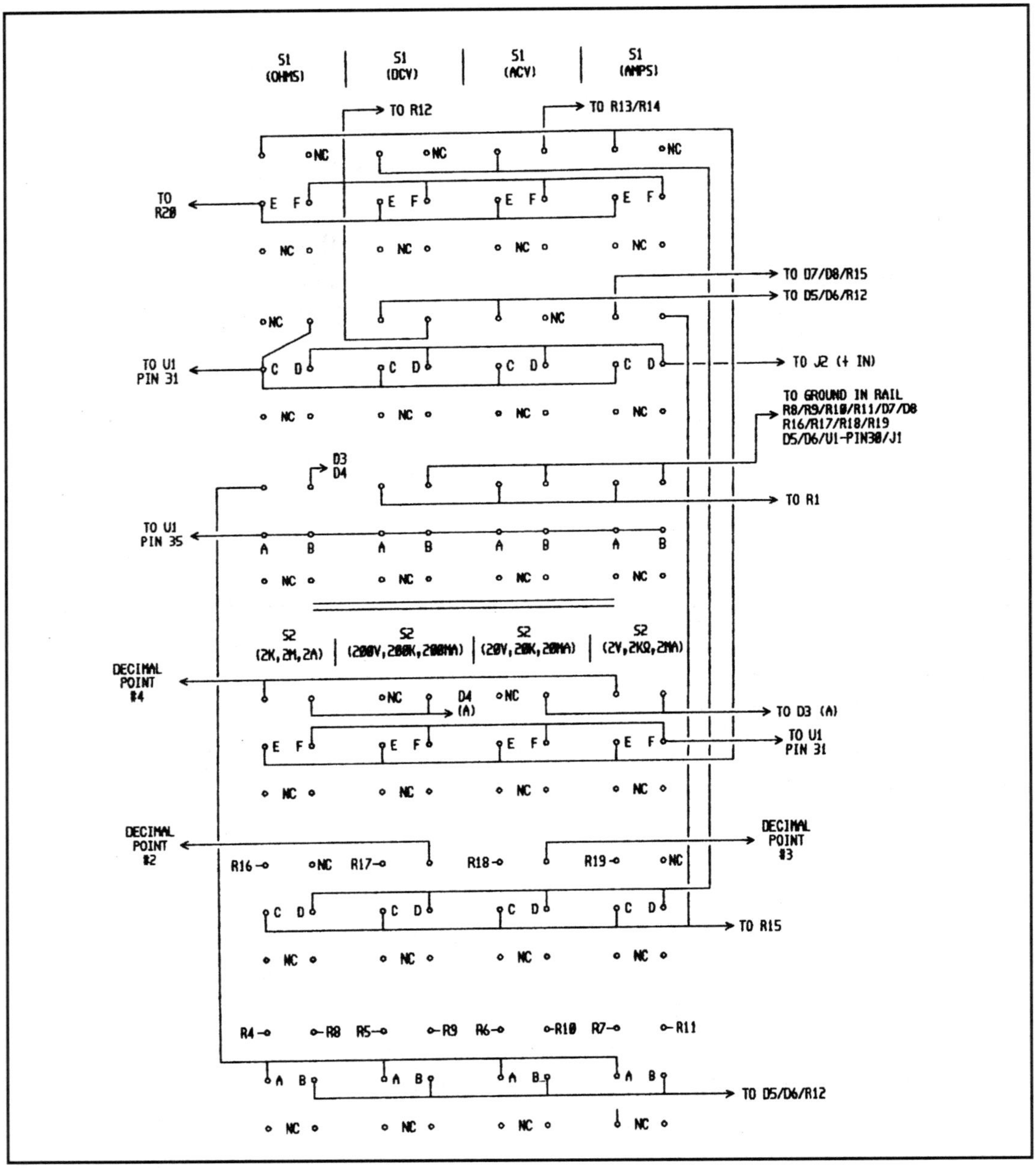

Figure 9-2. Multimeter switch connections.

circuit. Now, the value of any resistor placed across the input jacks will be displayed by the meter.

The final unknown quantity we want to appraise is current. This is done by placing a resistor, known as a *shunt*, across the inputs. The principle is simple. A voltage characteristic of the anonymous current is developed across the shunt, then fed to the meter's input. Hence, the 7107 provides a digital display of the current.

As seen in *Figure 9-1*, switch S2C, working in conjunction with switches S1C and S1D, provides the shunt (one of resistors R16, R17, R18 and R19). Through this network, the shunt is applied across the inputs and does its job.

For AC voltages, the configuration consisting of capacitors C6 and C7, resistors R13 and R14, and diodes D1 and D2, provide rectification of the incoming alternating current. Once this is accomplished, the voltage can be applied to the meter as with any other DC potential.

Lastly, switches S1E, S2D and S2E, along with resistor R20, accommodate the placement of the decimal point (DP) for the various range settings. Since we are dealing with a common-anode display, taking any decimal point low, through dropping resistor R20, results in its illumination.

The switches merely decide which of three decimal points (D2, D3 or D4) needs to be lit. In this fashion, the 2-volt scale displays as *2.000*, 200 milliamps are displayed as *200.0*, and so forth.

Switch S3 is for display testing. When pin 37 of the 7107 is pulled high (positive 5 volts), all display elements will illuminate. This results is a *-1888* annunciation, indicating that the meter section is functioning properly.

The pushbutton switches I employed come in three-switch arrays, so I had one switch left over. The leftover unit became *test*, but this is an optional feature and can be left out if desired.

The last section is illustrated in *Figure 9-3*. It is the dual or *split* power supply necessary for the 7107 chip. As usual, I implemented a stan-

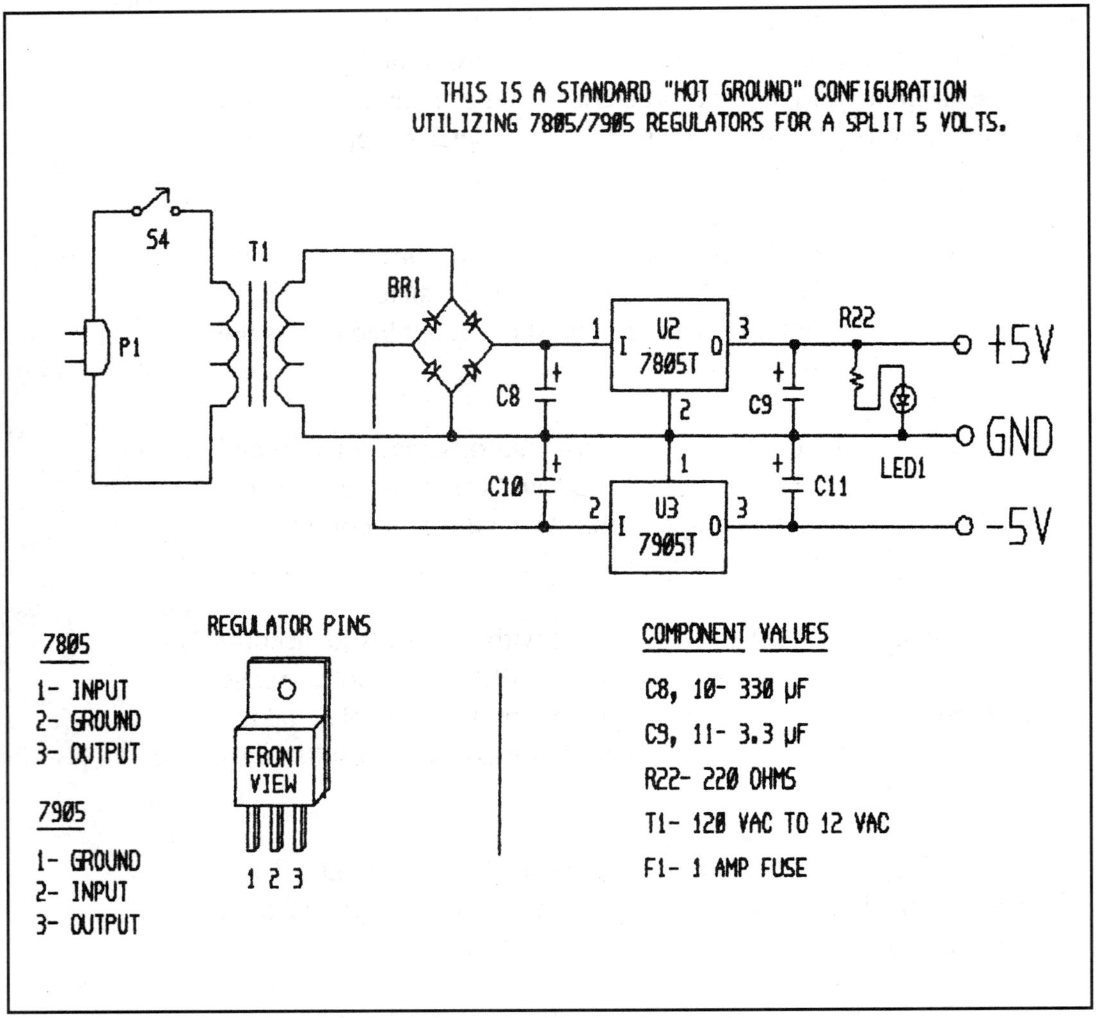

Figure 9-3. Multimeter power supply.

dard *hot ground* linear regulated system. *Hot ground* means that the ground rail attaches to one side of the AC input instead of to a transformer center tap.

Bridge BR1 handles rectification, while capacitors C8 and C10 filter both sides of the initial DC. Regulators U2 and U3 furnish positive and negative 5-volt supplies, respectively, and C9 and C11 filter transient noise and prevent regulator oscillation.

Resistor R22 and LED1 are wired across the positive 5-volt output, as an *on* indicator, completing this simple circuit with the requisite +/- 5 volts.

That about covers the operation of this system. Thanks to the 7107, tackling the theory of this multimeter is a fairly easy task. As you will see in the next section, the most problematic aspect of the project is wiring the switch assembly. So, let's move on to the fun stuff: constructing our digital bench multimeter.

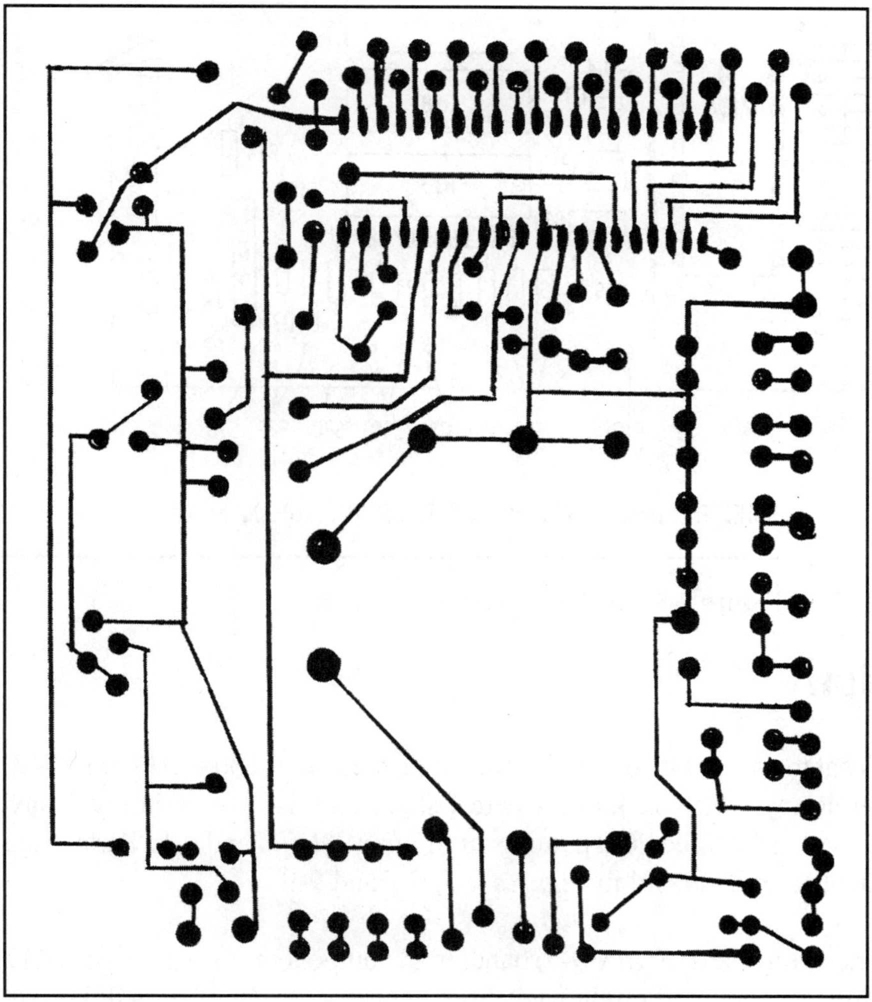

Figure 9-4. Multimeter main board foil.

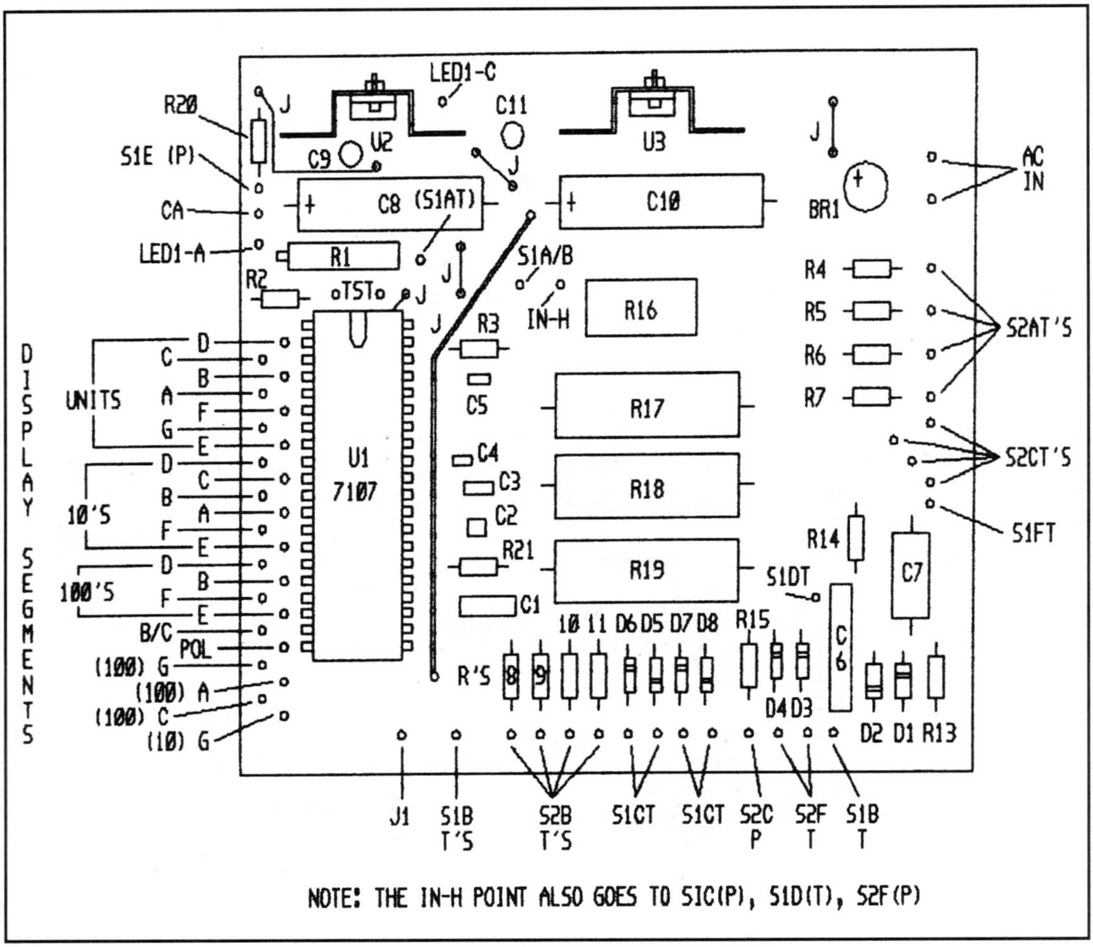

Figure 9-5. Main board parts placement.

Construction

What to use, point-to-point wiring or printed circuit board (PCB)? Well, by now you should know where I stand on this. Accordingly, I have provided the three PCB patterns in *Figures 9-4*, *9-6* and *9-8*. Parts placements are illustrated in *Figures 9-5*, *9-7* and *9-9*.

The main board (*Figure 9-5*) handles all components except resistor R12, which is mounted on the switchboard (*Figure 9-7*). This will be covered later.

The first step is to install the six jumpers and the 40-pin IC socket (per *Figure 9-5*). Next comes the small resistors, capacitors and diodes. Potentiometer R1 is next, followed by the four *current-measuring* resistors, R16, R17, R18 and R19. These resistors are 10-watt *power* units. Current measurements can generate some wattage; however, the 10-watt size is a little excessive. I used this size primarily because of availability, but smaller 2-, 3- or 5-watt components will do as well. Also, the 0.1-ohm and 1-ohm values of R16 and R17 are not always carried by distributors as standard items. See the parts list for dealers that carry these values. Being power resistors, they do take up some room. The 10-watt units measure 1-7/8" (48 mm.) by 3/8" (10 mm.) square, and space for them is allocated in the center right section of the main board.

Install the large power supply capacitors and the regulator/heat sink assemblies. This will complete the board, except for the hookup leads.

The next board (*Figure 9-7*) holds the pushbutton switches and resistor R12. Since R12 runs between the poles of switches S1F and S2B,

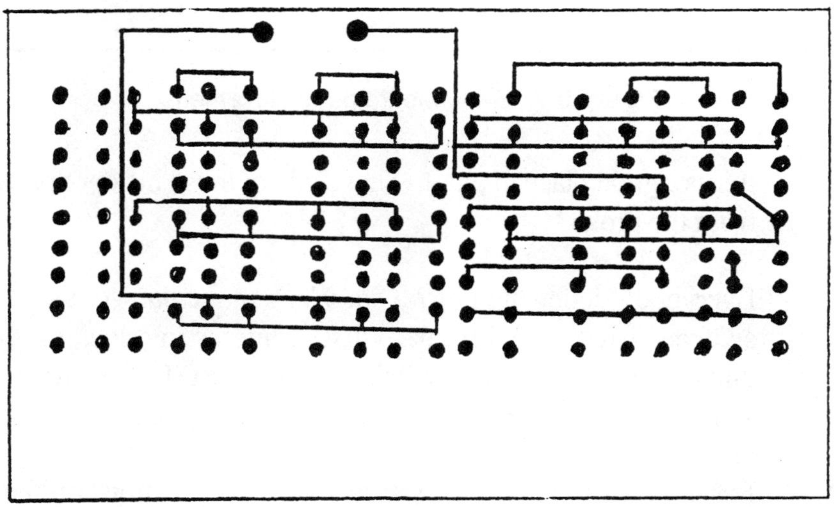

Figure 9-6. Switchboard foil pattern.

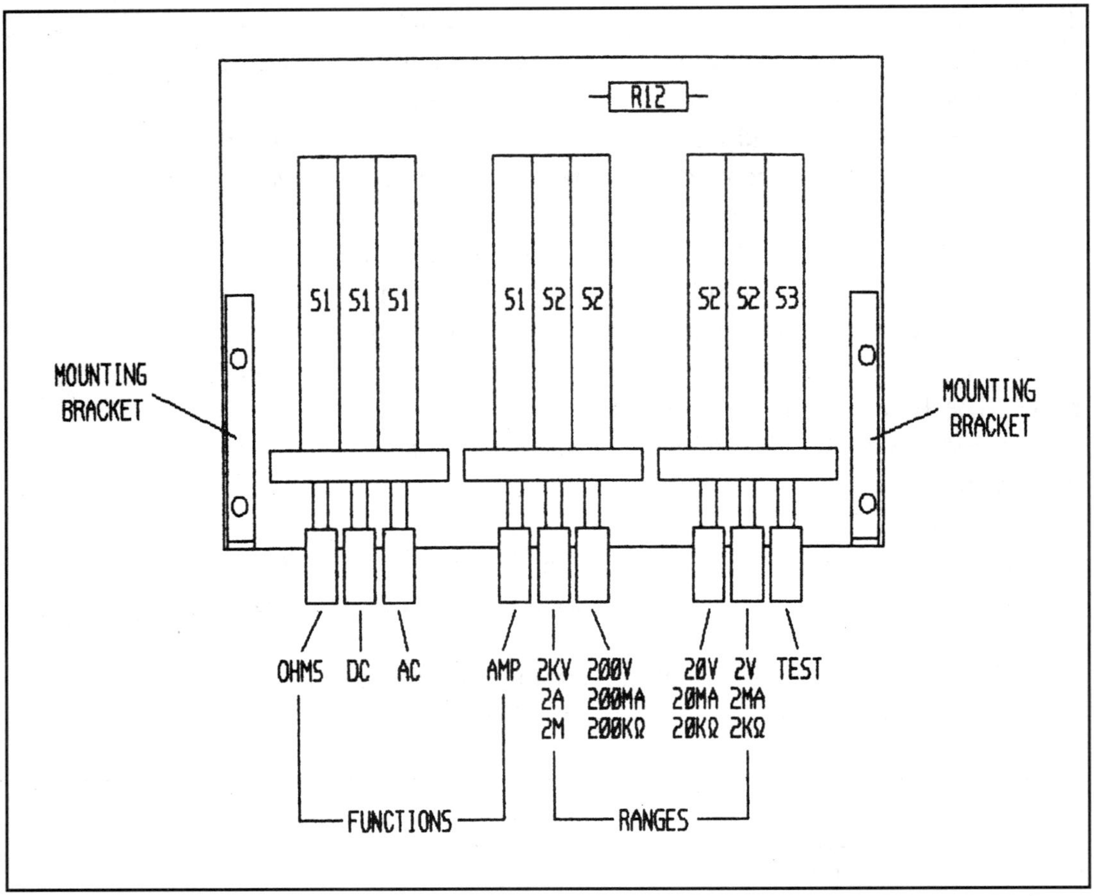

Figure 9-7. Switchboard parts placement.

this is a good place to put it. That's one of those insider tips you get only from the pros.

Everybody done laughing? Okay, let's move on. As I said before, the switches I used (see the parts list) come as interlocking three-switch units. Each individual switch is of the 6PDT persuasion; thus, four switches are required for both the S1s and the S2s.

For greater clarity, I used rotary switch designations on the schematic diagram (*Figure 9-1*). But as *Figure 9-2* illustrates, the six poles of each switch correspond with the *A-F* poles on the schematic. Only one set of throws is employed, so four switches handle the four throws of the designated schematic switches.

If this seems confusing it's because it is. Switches! They can drive you to the big house in a hurry. So, let me try to explain this in a practical sense.

You'll note that there are two series of switches; the S1s and the S2s. For the S1s, the first position is the resistance or *ohm* setting. However, there are six S1s (A, B, C, D, E and F); so the first pushbutton switch (designated *ohms*) handles all six. In the same fashion, the second pushbutton (designated *DC volts*) manages the second set of six positions. The third switch is for AC volts, and the last switch is for amperage.

The same principle applies to the *range* or S2 switches. Again, each individual switch represents one of the four *throw* positions, as shown in the schematic. The first S2 pushbutton is for 2 kilovolts, 2 megohms or 2 amps, depending on the S1 switch being activated, while the second handles 200 volts, 200 kilohms or 200 milliamps. The third S2 switch covers the next range, and the last one manages the smallest measurements (2 volts, 2 kilohms and 2 milliamps). Thus, by using eight 6PDT pushbutton switches, all four functions and all four ranges can be accessed.

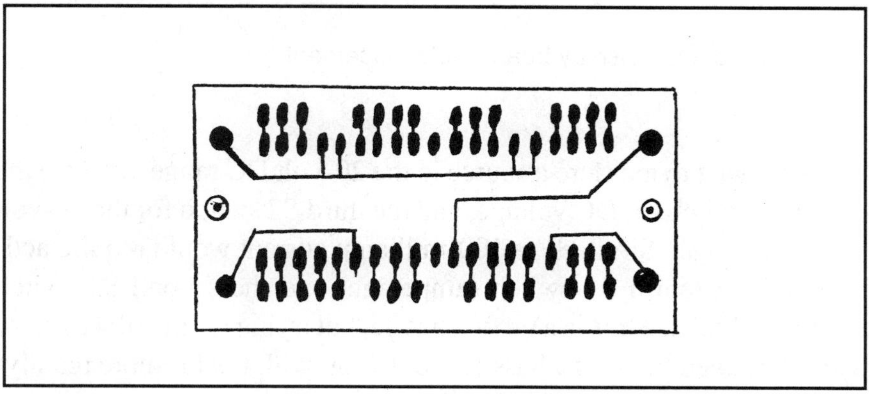

Figure 9-8. Display board foil pattern.

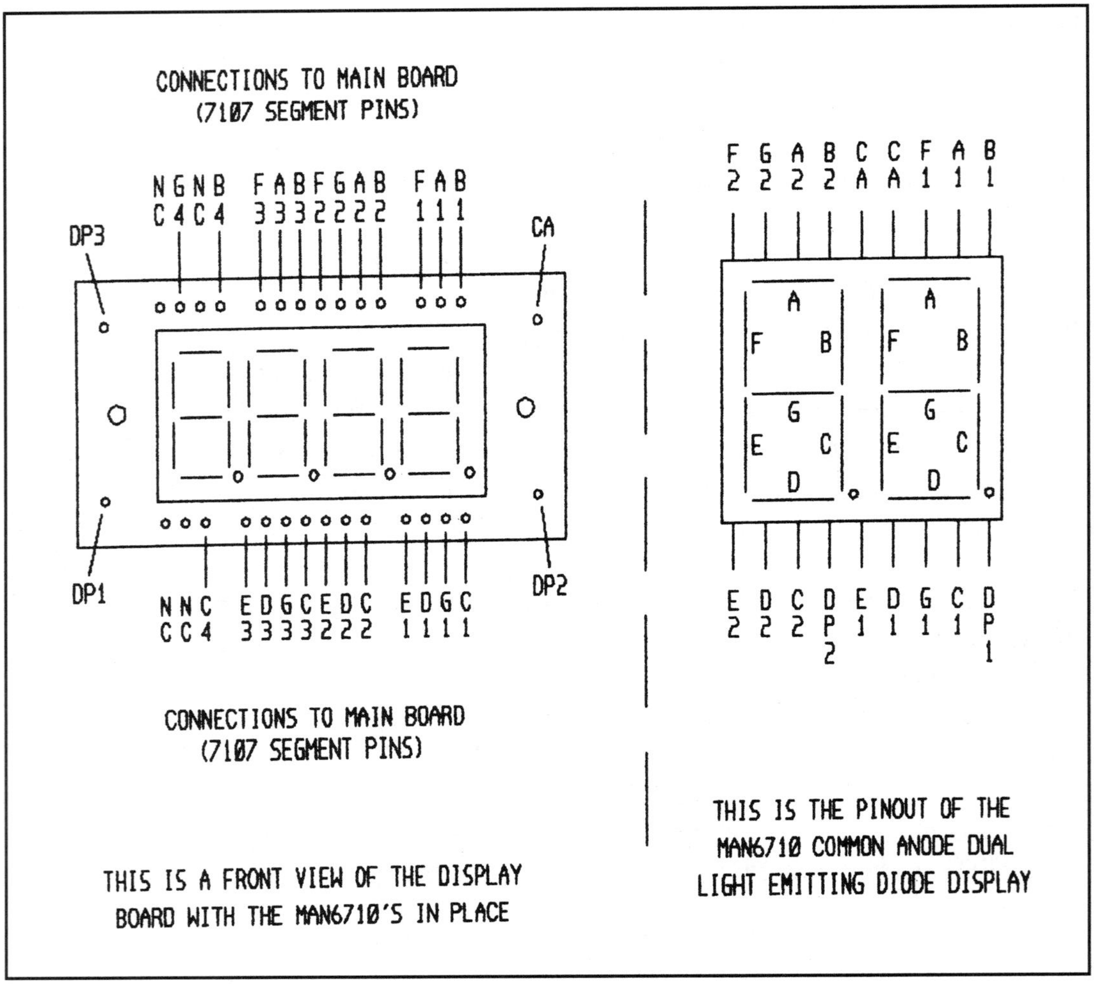

Figure 9-9. Display board parts placement.

So, if you want to measure a source in the 20-volt DC range, activate the second S1 switch for DC voltage, and the third S2 switch for the 20-volt maximum range. Similarly, a 100-milliamp current would require activation of the fourth S1 switch (amperage) and the second S2 switch (200 mA). I realize this is a bit of a hassle, but trying to find 6P4T rotary switches is even more of a hassle. So, I went with the far more readily-available pushbutton switch assemblies.

Hopefully you have the idea. But if you use the provided PCB pattern and wire the switch assembly as shown in *Figure 9-2*, it will work whether you understand it or not.

First, solder the three switch arrays to the PCB (*Figure 9-7*), then install resistor R12. Now, attach appropriate lengths of hookup wire to the various switch points, as seen in *Figure 9-2*. These will later be soldered to the main and display boards. It is a good idea to label each lead according to where it should go on the main board. This will help you during the final wiring.

With that done, set the switchboard aside and turn your attention to the display (*Figure 9-9*). First, install the 40-pin IC socket that will hold

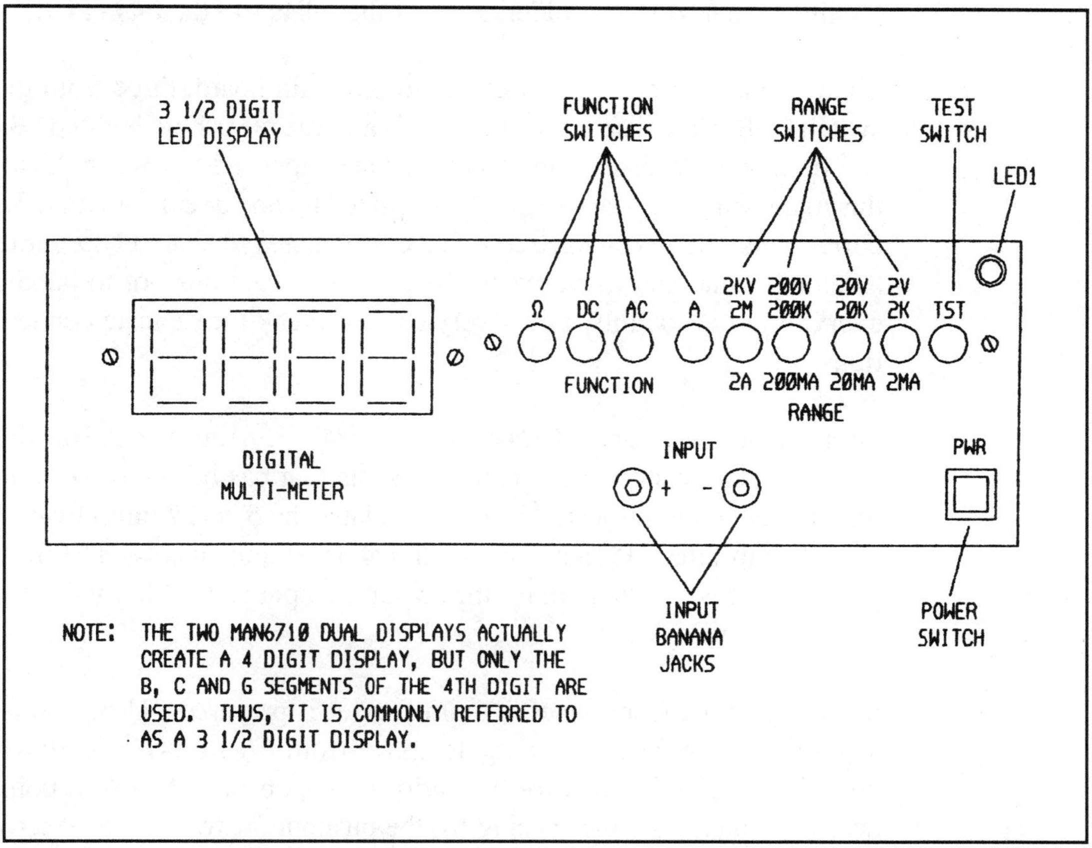

Figure 9-10. Multimeter front panel template.

the two dual displays. Then on the back side, run the jumper from the fourth *B* segment to the fourth *C* segment. This will be the *1* or *1/2* digit.

As for the dual displays, I used two MAN6710 common-anode 0.56" dual-LED display units. The pinout is shown in *Figure 9-9*; however, any common-anode display will work. Many of the large dual arrays have the same pinout as the 6710.

Now, attach appropriate lengths of hookup wire to the display board for connection to the main board. When I say "appropriate lengths," I merely mean lengths that will easily reach from the display panel location to the main circuit board positioned in the housing. This isn't rocket science; it's obviously better to have the hookups a little long than too short. Again, it helps to take the extra time to mark each lead. This will greatly facilitate the next step: soldering all of these leads to the main PCB.

There are a total of 54 connections to the main board, three from the switch PCB to the display, so patience is a virtue regarding this step. Be careful that all leads end up where they are supposed to end up. With this many wires, it's easy to get things mixed up and a real pain to make corrections. Once you have completed that chore, you can set the entire assembly aside and move on to the case. Be cautious not to handle everything too roughly, as it's very easy to break the delicate connections.

For the case, I used an 8" (204 mm.) by 6-1/4" (158 mm.) by 2-1/2" (63 mm.) standard instrument design; the same that has been employed in many of the other projects. It accommodates the 5" (127 mm.) by 4-1/4" (108 mm.) main PCB and power transformer quite nicely. The front panel provides ample room for the switches, input jacks, LED indicator and 0.56" displays.

Speaking of the front panel, it's time to get it prepared. Using drills, small files, a hobby saw and the template from *Figure 9-10*, make the holes in the plastic panel for the various components. A critical point here is to make sure the spacing for the pushbutton switches is exactly right.

I used round switch caps. Be careful when using these, though: if you get the hole off even slightly, the caps will rub against the plastic. This sometimes interferes with the switch action, and all in all, is not a good deal (you know, sloppy work). So, watch this procedure very carefully.

With that done, paint the sucker. I always use light yellow, as it makes it easy for me to read the black dry-transfer lettering. However, other light shades will work as well. Once the paint job has dried completely, label the front panel with the dry-transfer letters. *Figure 9-10* illustrates how I designated each function. Last, give the panel a light coat of clear enamel to protect your work.

The back panel requires only a single hole large enough to accommodate a rubber grommet and the power cord. Install the grommet, feed the cord through it, and tie a knot in the cord for strain relief (on the inside, of course). This completes the back panel. You could also paint it; but since it's the back of the unit, I never bother. As for the shell sections of the case, you don't need to do anything to the top half, including painting. As a matter of personal preference, I paint the shell a medium blue, but it's all up to you.

The bottom half needs only two holes to secure the power transformer. Again, I gave it a coat of blue paint, and that was that. Allow the paint to dry, then install the panel components. Hookup leads should already be connected to everything but the banana jacks, since they have to go through the panel before soldering. Make these connections and secure everything in place.

As for the switchboard, I used small, homemade brackets to attach it to the front panel. You might have a better idea. Resistor R22, the *on* LED dropping resistor, is soldered directly to LED1. This saves putting it on the main board.

At this point you're almost done, but don't put everything in the shell just yet (*Photo 9-1*). The reference voltage has to be set with potentiometer R1, and its position in the prototype dictates this be done out-

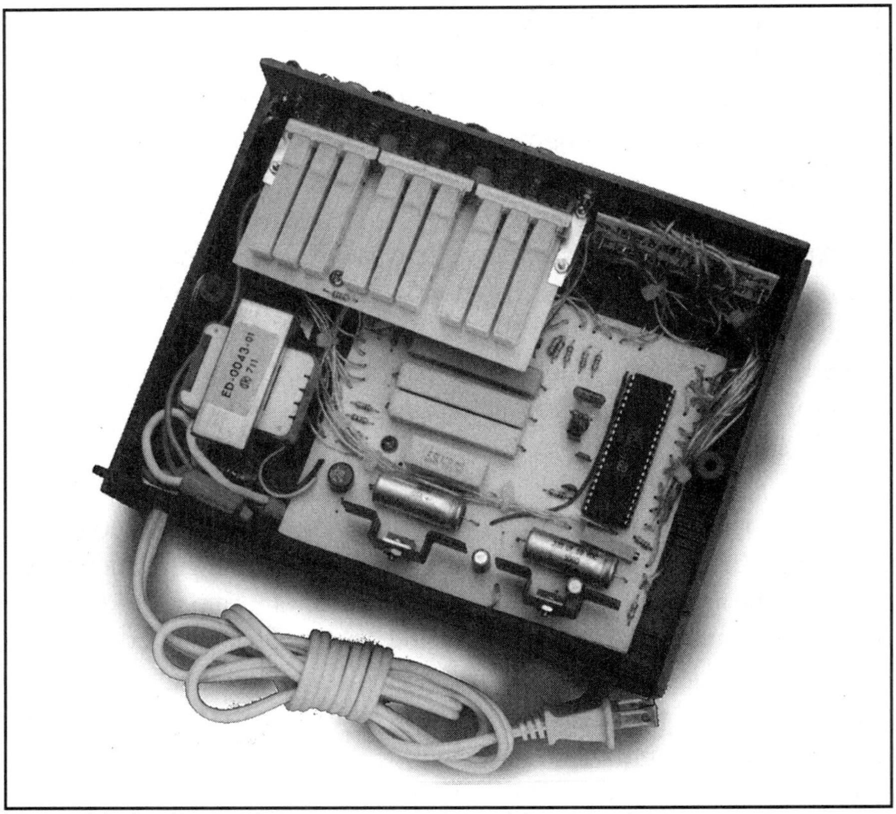

Photo 9-1. Internal view of the completed bench multimeter. Note how the switchboard mounts above part of the main board.

side the case. Since this is part of the next phase, let's move on to the testing/calibration stage.

Testing, Calibration and Operation

Calibrating the digital multimeter is simple through the adjustment of R1, but you can't do this until the system is working. So, hit the power switch (S4), and you should see either *000* or *1—*, depending on the function selected. If you don't see either of these, check the power supply for +/- 5 volts. If this isn't the problem, double-check your wiring. This is a relatively simple circuit, so errors should be easy to spot.

If all seems well, press the test button, and the display should read -*1888*. Now select the DC function and the 20-volt range; the second S1 switch and the third S2 switch. Without any input, the display will read *000*. Now attach a 9-volt battery to the input. The display should give a numerical reading.

At this point, the reading will probably not be 9 volts or anywhere near it. This is because we haven't yet calibrated the 7107 circuit. For this, another DMM (digital multimeter) is required. A portable unit will do fine.

First, take a reading from the battery to determine the exact voltage provided. Now, connect the battery to the meter's input and adjust R1 until the display matches the initial reading. Using a 10-turn style potentiometer for R1 makes this process easier and far more accurate.

Next, try applying several different DC sources to the input. In each case, first check the actual output of the source before connecting it to the bench meter. It is highly likely that you will see almost identical readings, but you may have to make a slight adjustment to R1. Strive for a happy medium.

Another calibration method is to measure the reference voltage at pins 35 and 36 of the 7107. Since a properly set reference voltage is half the full-scale reading of the meter, and since we have configured the meter to read 200 millivolts full scale, the measurement should be 100 millivolts. Again, adjusting R1 will bring this voltage into line.

Once you are satisfied that the bench meter is as accurate as possible, the calibration is completed. Your meter should now reliably measure resistance from 1 to 2,000,000 ohms, AC and DC voltage from 1 to 2,000 volts and amperage from .001 milliamps to 2 amps.

Try setting the meter for the 200 kilohm range (first S1 switch/second S2 switch). Begin testing various resistors of known value to see how close they come to their advertised value. Also, using the independent

DMM, compare the two readings. Again, I suspect you will find them very close.

Anyway, what you need to remember is that to take a measurement, you have to activate one of the function switches, and one of the range switches.

These are interconnected switch assemblies, which means that pressing any one of the four switches automatically releases any previously activated switch. Remember, we have two banks of four switches each.

Suppose you want to use the 2-amp settings? Both switches are in the same bank, and you will release the first switch when you press the second one, right? Well, this is theoretically true, except that if you press both switches at the same time, they both stay activated. Another one of those insider tips. This, of course, is true for any combination employing two switches in the same bank. Also, it is worth noting that a switch is only activated when it is depressed. This is because we did not make use of half the throws on our 6PDT units, which is the half that activates when the switch is released.

Incidentally, the 2-kilohm range can be used to check diodes. With the anode connected to J2 and the cathode to J1, a display of around *700* will appear. Reverse the leads and the display stays at overflow (*1*—).

Your system is working, calibrated, and you know how to use it, right? Now, when you need a DMM at the bench, you won't have to fiddle around with small portable devices. This heavyweight will stay in its place and out of the way.

Conclusion

In addition to an ample supply of components, a well-stocked electronics workbench needs many different instruments. One of the most important (if not essential) items is a multimeter (*Photo 9-2*). It doesn't necessarily have to be digital, but it helps. For that matter, the DMM is

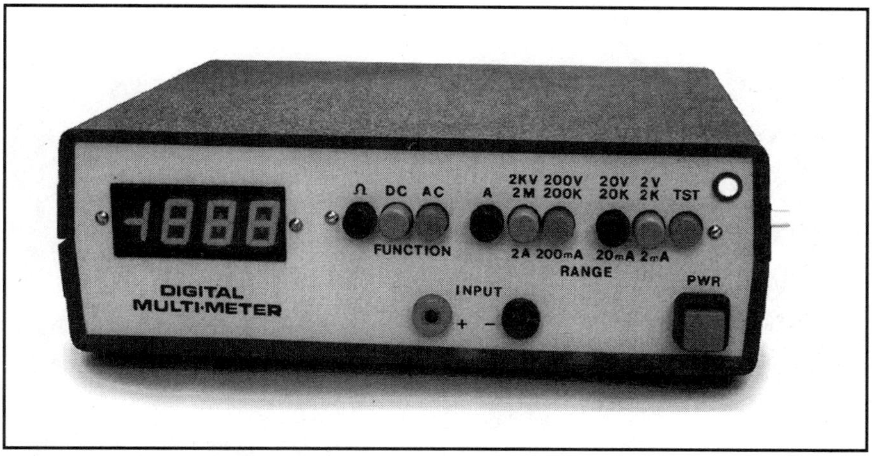

Photo 9-2. The completed bench multimeter. The display has all of its segments lit, which is the result of activating the *test* mode.

probably the best place to start when you first set up your lab. Few pieces of test equipment offer as many functions or as much versatility as the multimeter. And, with the addition of low-cost adapters, such as the CA-1 we saw in Chapter 7, the DMM becomes an even more versatile tool.

Hence, if you build nothing else in this book, I strongly recommend you do try this project. It will quickly prove its merit, and if all else fails, you can use it to monitor line voltage. It is always nice to know about those brownouts. Additionally, this project also presents you with a chance to work with the ICL7107 integrated circuit. This, in itself, makes the project well worth the effort. Get to know this chip. You won't regret it.

Chapter 10
The Digital Function Generator

Introduction

How often have you needed a source of stable sine, triangle or square waves? I imagine the answer is probably "many times." With today's integrated circuits and other technologies, solid signals of all three waveforms are frequently invaluable for prototypes and circuit designs.

This is where the function generator comes into play. Utilizing a single voltage-controlled oscillator (VCO) chip, this device will produce all three waves simultaneously, and do so at the frequency you select, up to about 2 megahertz. Thus, when a sine (sinusoidal), triangle (sawtooth) or square wave is needed, this remarkable chip is at your service.

The addition of a simple frequency counter provides an accurate digital display of the output frequency, eliminating the need for a bulky, less-reliable dial-type tuning indicator.

An extra feature of the counter is that it can be accessed separately and used as a second or third, or whatever, bench counter. I like this function, as there is always room for another frequency counter. Well, almost always. So, when you complete this project, you will not only have a highly stable signal generator, but a spare counter as well. I think you'll find both functions welcome additions to your bench, especially since they come in a single package.

DIGITAL FUNCTION GENERATOR PARTS LIST

SEMICONDUCTORS:

U1-4	CD4026 Decade Counter/Dividers (JDR Micro)
U5	CD4047 Monostable/Astable Multivibrator
U6	ICL8083 VCO/Precision Waveform Generator
U7	74HC4049 Hex Inverter
U8	LM386N-3 Low Power Audio Amplifier (9V/500mw)
U9	7812T Positive 12 Volt Regulator
BR1	400 PIV 1 1/2 Amp Bridge Rectifier (WO4M, or Equiv)
DSP1-4	2-MAN6740 Common Cathode Dual LED Displays
LED1	Green T-1 3/4 Jumbo Light Emitting Diode

RESISTORS

R1, 5	1,000 Ohm 1/4 Watt Resistors
R2	1,000,000 Ohm PCB Potentiometer
R3	10,000 Ohm 1/4 Watt Resistors
R4	10,000 Ohm 10-Turn Panel Potentiometer (Electronic Goldmine, Circuit Specialists)
R6, 8	100,000 Ohm 1/4 Watt Resistors
R7	100,000 Ohm Panel Potentiometer
R9	10,000,000 Ohm 1/4 Watt Resistor
R10	10,000 Ohm Panel Potentiometer W/Switch
R11	2,200 Ohm 1/4 Watt Resistor

CAPACITORS

C1, 6	0.01 Microfarad Mylar Capacitors
C2, 5	0.001 Microfarad Mylar Capacitors
C3	200 Picofarad Mica Capacitor
C4	100 Picofarad Mica Capacitor
C7	0.1 Microfarad Mylar Capacitor
C8	47 Microfarad Electrolytic Capacitor
C9	470 Microfarad Electrolytic Capacitor
C10	0.22 Microfarad Monolythic Capacitor
C11	220 Microfarad Electrolytic Capacitor
C12	3,300 Microfarad Electrolytic Capacitor

OTHER COMPONENTS

S1	SPST Pushbutton Switch
S2, 6	DPDT Slide Switches
S3, 4	2P6T Rotary Switches
S5	SPDT Slide Switch
S7	SPST Switch on Potentiometer R11
J1-4	Banana Jacks
P1	120 VAC Line Cord and Plug
T1	120 VAC to 12.6 VAC Power Transformer
T2	8 Ohm to 1,200 Ohm Audio Transformer
SPKR	1 to 1 1/2 Inch 8 Ohm Speaker

MISCELLANEOUS

PCB Materials, Solder, Hook-Up Wire, LED Holder, IC Sockets, Heat Sink, Paint, Dry Transfer Lettering, Knobs, Hardware, Clear Red Plastic, 8" by 6 1/2" by 2 1/2 " Instrument Case, Etc.

Theory

Intersil/Harris comes to the rescue for this venture. Their ICL8038CCPD VCO/precision waveform generator is the ideal starting place. As previously stated, it will produce all three basic waves at once, all at the preselected frequency of your choice.

Figure 10-1 illustrates the 8038 block diagram, revealing a very clever design. A constant current source is used to charge an external capacitor to two thirds the operating voltage (VCC). As it reaches that level, one of the comparators activates the flip-flop circuit, allowing the capacitor to start discharging. When the discharge is down to one third VCC, the other comparator resets the flip-flop, and the charging cycle starts again. This charge/discharge activity produces oscillation governed by the external capacitor.

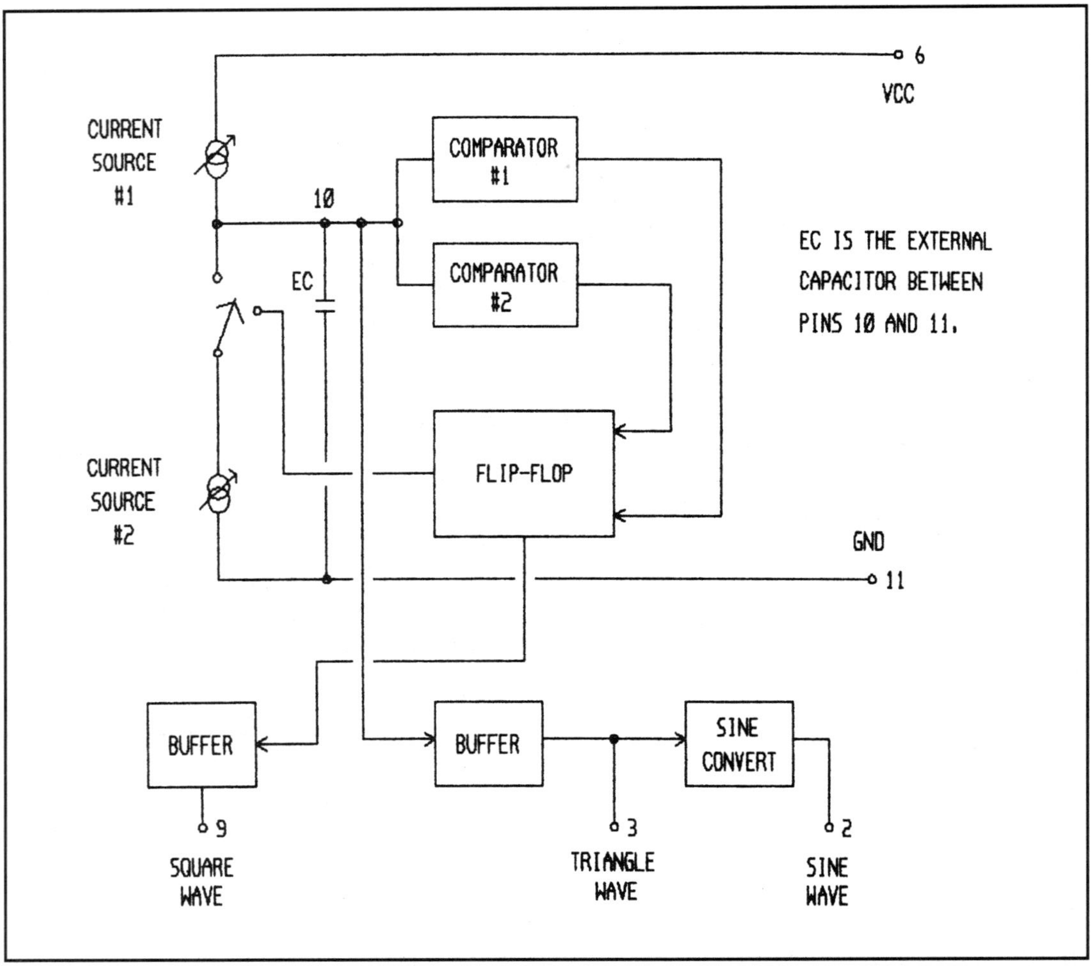

Figure 10-1. ICL8038 block diagram.

If this sounds a lot like the operation of the LM555 timer, give yourself a gold star. It is a lot like that; however, the 8038 takes it a little further. The flip-flop output provides the square wave, while the triangular signal is the result of the constant current sources charging and discharging the capacitor. Both are fed to buffers which furnish the respective outputs. The sine wave is created by sending part of the triangle signal to a *sine converter*. This is nothing more than a circuit that rounds the tops off of the triangle, allowing it to appear as a sine wave. This is not a pure signal, but as stated, it's good enough. Intersil data sheets will tell you that the range is from 0.001 hertz to 1 megahertz, but I have had no

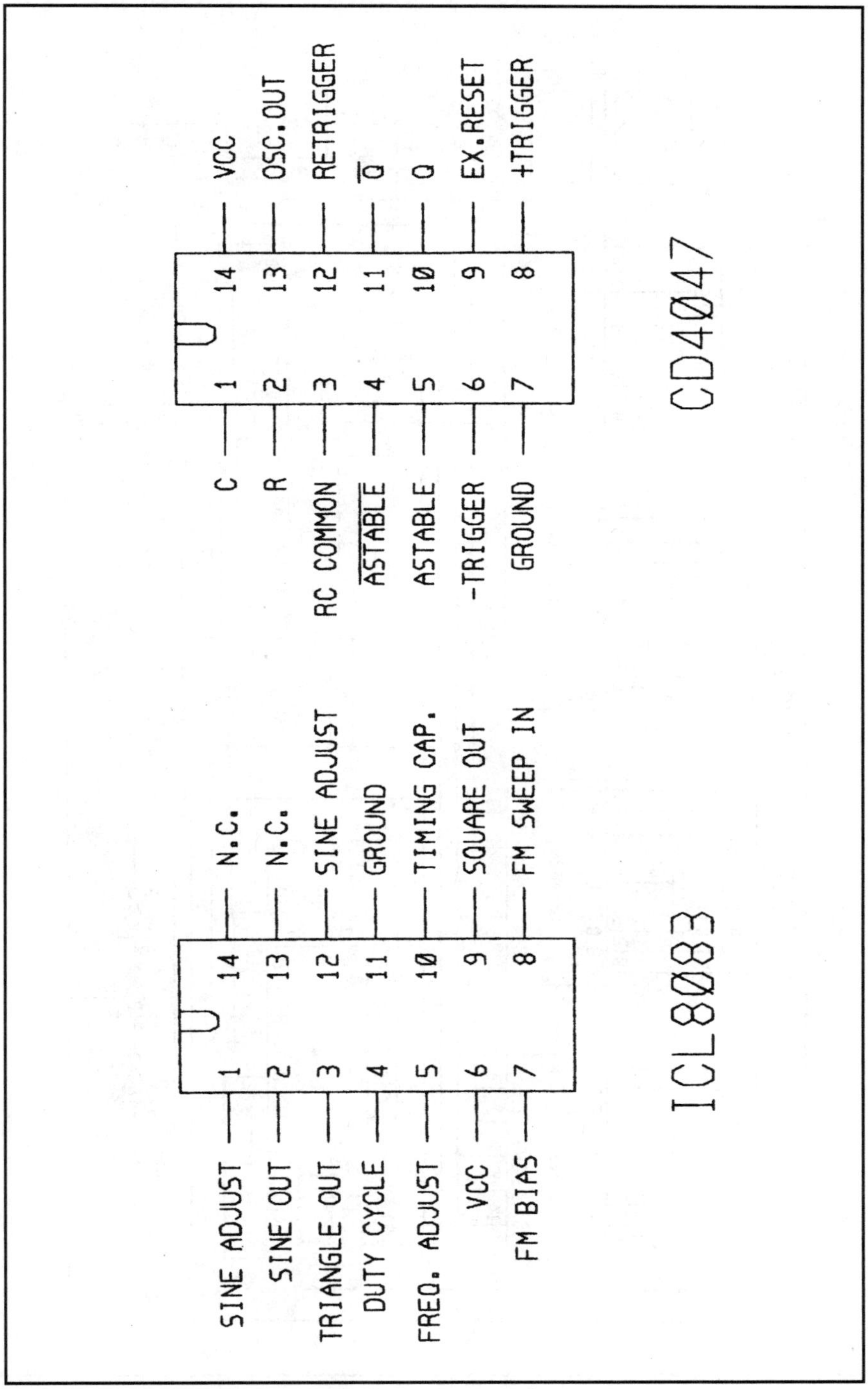

Figure 10-2. ICL8038 and CD4047 pinouts.

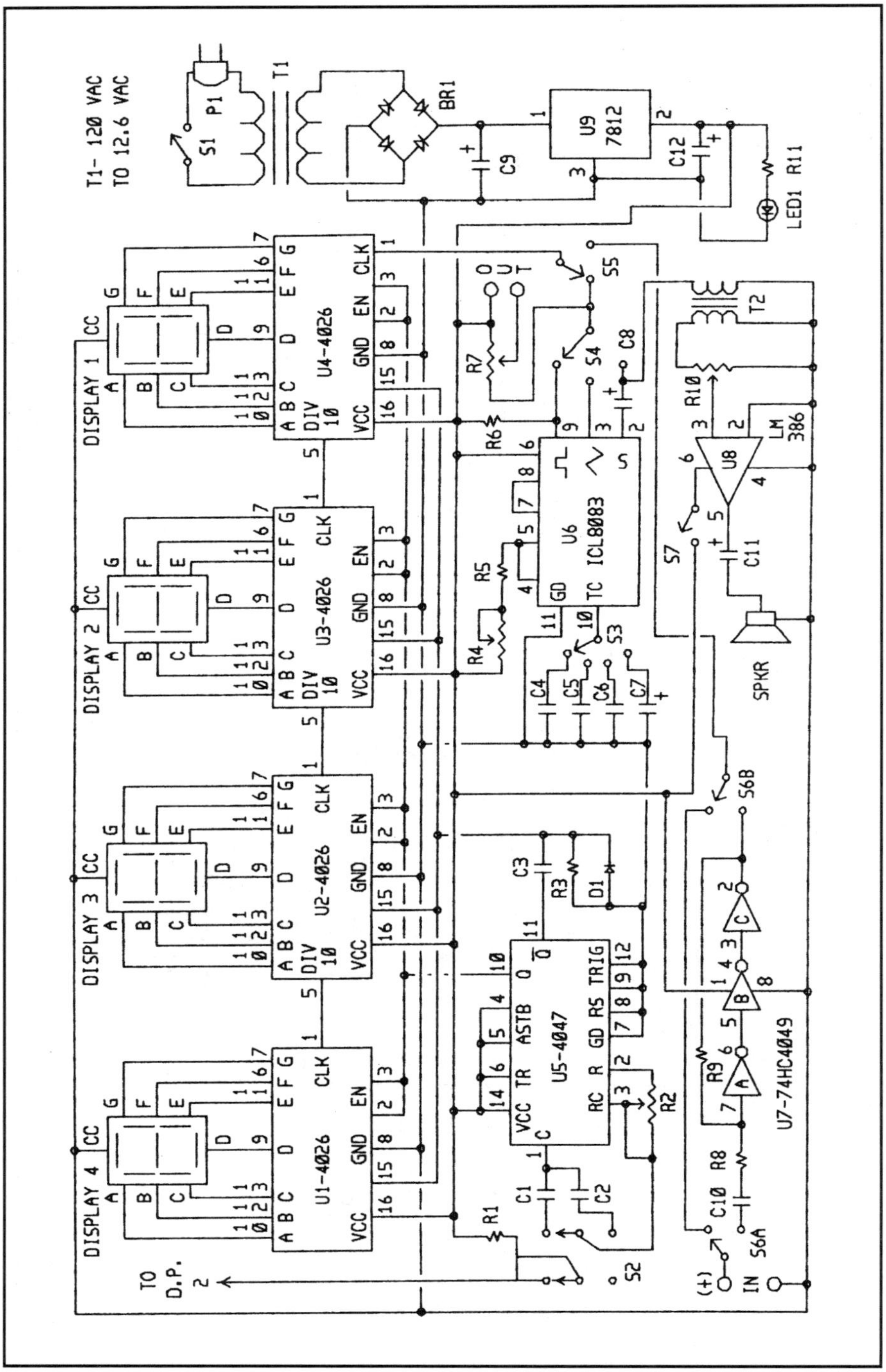

Figure 10-3. Digital function generator.

trouble getting 2 megahertz out of this design. This easily covers the audio range and a useful RF test spectrum.

Outputs are as follows: 1 volt peak-to-peak (p-p) for the sine wave, 2 volts p-p for the square wave, and 4 volts p-p for triangular signals. This is more than enough for most situations. A quick note: these levels are for operation with a 12-volt power supply. The 8038 will function with a VCC as low as 6 volts, but the outputs will be proportionally lower.

All of this superb circuitry comes in a nifty 14-pin dual in-line pin (DIP) package that is very convenient to use. *Figure 10-2* provides pinouts for the 8038 as well as for the other chips used in this project. With the addition of a handful of external components, you will quickly have a highly-reliable function generator.

So much for the basic operation. Let's take a look at the 8038's role in the full design.

As seen in *Figure 10-3*, one of four external capacitors is selected by switch S3 and connected to pin 10 of our VCO chip. This, in conjunction with the resistor/potentiometer network between pins 4 and 5, establishes the operational range. R4 is the fine-tuning control, and for the prototype, I used a 10-turn potentiometer. This is not essential, but it does make the tuning process more precise.

Switch S4 selects the desired waveform, and potentiometer R7 handles the output intensity (gain). Also connected to the pole of S4 is one throw of switch S5. The other throw goes to switch S6B, but I'll get to that later.

The pole of S5 is routed to the clock input of integrated circuit (IC) U4. U4 is part of the frequency counter circuit. In this case, the counter is made up of five chips. The first four, U1-4, are CD4026 decade counter/dividers, while the fifth is a CD4047 monostable/astable multivibrator. As with all frequency counters, this one takes the incoming signal, divides it by 10, then sends the product to the next stage and displays the remainder.

The 4026 has the very handy property of being able to directly drive instrument-sized (0.56"), common-cathode light-emitting diode (LED) displays. Thus, external driver circuits are not required. However, to make this counting/dividing system a true frequency counter, some form of gate or window timing is necessary. You guessed it; enter the 4047, stage right.

As a multivibrator, the 4047 (U5) can employ a resistor/capacitor (RC) tank circuit to establish the gate; which is exactly what we will use it for. Switch S2 selects one of two capacitor values, giving the system two window times and two count resolutions.

Naturally, the R/C approach will not produce the timing accuracy of a crystal-controlled system. However, for our purposes, the tank is more than satisfactory. If greater signal measuring precision is required, the frequency counter in the next chapter will cheerfully provide it.

The 4047 output is sent to the enable pins of the 4026 chips, so the four ICs function only during the gate time. Thus, the display will accurately annunciate frequencies between 100 hertz and 1 megahertz, depending on various other settings.

In review, during the gate time, the incoming signal is divided by four stages. The remainders of each division appear on the four-digit display. For example, say our signal is 125 hertz. The first stage will divide 125 by 10 for a product of 12 and a displayed remainder of *5*. The second stage will divide 12 by 10 with the product of 10 and the remainder displayed as *2*. The third stage will divide 10 by 10 and display a *1*. As you've run out of numbers, the fourth stage is not involved.

Now, back to switch S5. This switch selects either the *output* of the 8038 generator or an *input* from an external signal source. Whichever is chosen is then sent to the clock input (pin 1) of the first counting/dividing stage (U4).

Since our frequency counter (like most) is digital, it only recognizes square/triangle wave signals. So, if you want to introduce a sine wave, it has to be conditioned before the counter can read it. That's the job of the 74HC4049 IC (U7). Input conditioning circuits range from complex operational amplifiers (op-amps) to linear *choppers*, and are the subject of great debate. Both have their good and bad sides. Op-amps produce respectable results, but usually require split (+/-) power supplies. Chopping circuits, on the other hand, run on single supplies, but are often frequency limited. So, the ideal conditioner remains a matter of heated argument.

The circuit I selected for this project first came to my attention in an article by Mark Emery Bolles (Gernsback publications). Mark built a similar frequency counter and used this configuration for conditioning. I liked its simplicity, tried it out, and it worked beautifully. To quote Mark from his article, "I am not above stooping to theft (although I prefer to call it 'research')." My sentiments exactly. So, here is Mark's circuit, slightly modified.

The principle behind this signal conditioner is analogous to the 8038's sine wave converter. In this case, however, the 74HC4049 chops off the tops, thus simulating a square wave. You may ask, "How is this done?" Well, hold on, I'm gonna tell you.

The trick here is to employ the 74HC4049 in a linear fashion, as opposed to its traditional *logic* role. By selecting resistors R8 and R9 to produce a gain of 100 (10 megohm/100 kilohm = 100), even a small input signal will saturate the HC4049. This in turn will provide fast rise and fall times, and flat tops. That's right: you'll have a square wave. It's simple, but effective and clever. This is one aspect of electronics that never ceases to amaze me: the versatility of so many of the logic series and linear ICs. With few exceptions, far more can be done with these chips beyond their original, intended applications.

So, there you have the input conditioning. Switch S6 allows you to cut this circuit in or out of the input rail as the circumstance dictates. Now, the internal frequency counter can handle all three waveform types, instead of just square and triangle signals. You'll find this really handy when it comes time to measure radio frequency (RF) and audio signals. The procedure will crop up more often than you might expect.

Integrated circuit U8, an LM386 low-power audio amp, is arranged in its most basic configuration. Normally, to improve tone quality, I add at least a 10-ohm resistor and 0.05 microfarad capacitor, in series, from the output to ground. But for our purposes, this isn't necessary, so why use the extra parts?

Here, audio transformer T2 matches the 8038's sine wave output to the 386 input. This provides the ability to monitor the audio frequency range. The feature can be beneficial for a variety of reasons. It allows you to use the generator as an audio tone source, furnishes a sound signal when your attention is diverted from the display, and aids anyone who is sight disabled. Potentiometer R10 acts as a volume control, while switch S7 permits you to turn off the amp when audio is not desired. To conserve panel space, the prototype incorporates S7 on the potentiometer.

The last section is the power supply. Here we need a regulated 12 volts, so let's use our old friend the linear supply. As usual, transformer T1 supplies the AC, bridge BR1 handles rectification, and capacitor C9 filters out the ripple. A 7812T IC (U9) does the regulation honors, and capacitor C12 serves to filter transients and prevent the oscillation of U9. One marked difference with this supply is the value of C12. This is usually a 1- to 10-microfarad electrolytic; but in this design, a 3,300-microfarad unit is employed.

The large size is necessary for additional ripple filtering. This wouldn't normally be a problem, except for the audio amp. Without the big capacitor, you will hear a heavy 60-hertz hum while you monitor the audio frequencies.

Switch S1 is the power (on/off) control, and the LED1/R11 network provides *on* indication. It probably goes without saying, but I'll say it anyway: plug P1 is very important. It's used to extract 120 VAC power from the wall jack. In my experience, these generators rarely function properly unless plugged into the wall.

Construction

As usual, this unit can be built using either point-to-point wiring or a printed circuit board (PCB). I recommend the PCB approach, and have included the foil pattern in *Figure 10-4*. *Figure 10-5* shows the pinouts for the LM386 and 7812T. Also, *Figure 10-6* provides a parts placement diagram.

First install the jumpers, and note that three of these run partially under the 4026 (U1-4) IC sockets. The sockets come next, followed by the discrete components, bridge and regulator/heat sink arrangement. Be sure to solder the *shield* tabs of audio transformer T2 to the provided pads. It can cause some noise problems if not grounded. Next come the hookup leads. There are a bunch of them, so take care, especially with the display connections. If you have any *rainbow* ribbon cable on hand, the various colors will help you identify what goes where.

Since the speaker is mounted on the top half of the case, I chose to use a small header-style two-pin connector. It will allow you to disconnect the speaker whenever the case is opened, preventing the wires from being yanked loose from the main board.

At this point, a quick check of the power supply is in order. Hook up power transformer T1 and check for positive 12 VDC at pin 2 of the 7812 (*Figure 10-5*). You should see the same thing at the following socket locations: U1-4, pin 16; U5, pins 4, 5, 6 and 14; U6, pin 6; U7, pin 1; and U8, pin 6 (when switch S7 is closed).

If everything is satisfactory, the ICs can be installed and the panel components and display connected. Hold off on the power switch and banana jacks, as they have to first go through the front panel before you can solder the leads.

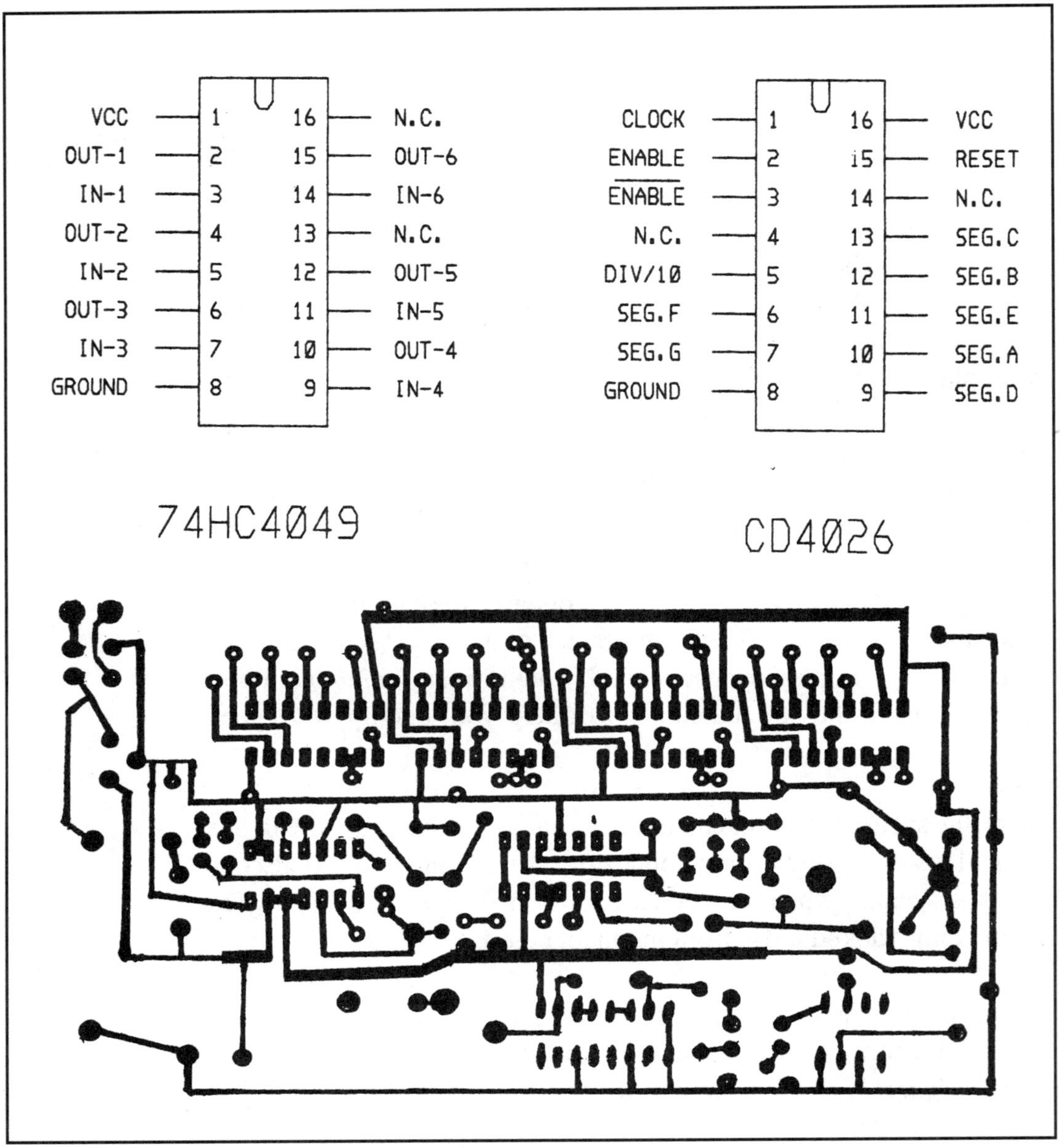

Figure 10-4. 74HC4049 and CD4026 pinouts, and function generator foil pattern.

Let's set the main board aside for the moment and turn our attention to the enclosure. The prototype is housed in a standard instrument-style case measuring 8" (200 mm.) by 6-1/2" (165 mm.) by 2-1/2" (63 mm.). The front panel is (to say the least) crowded.

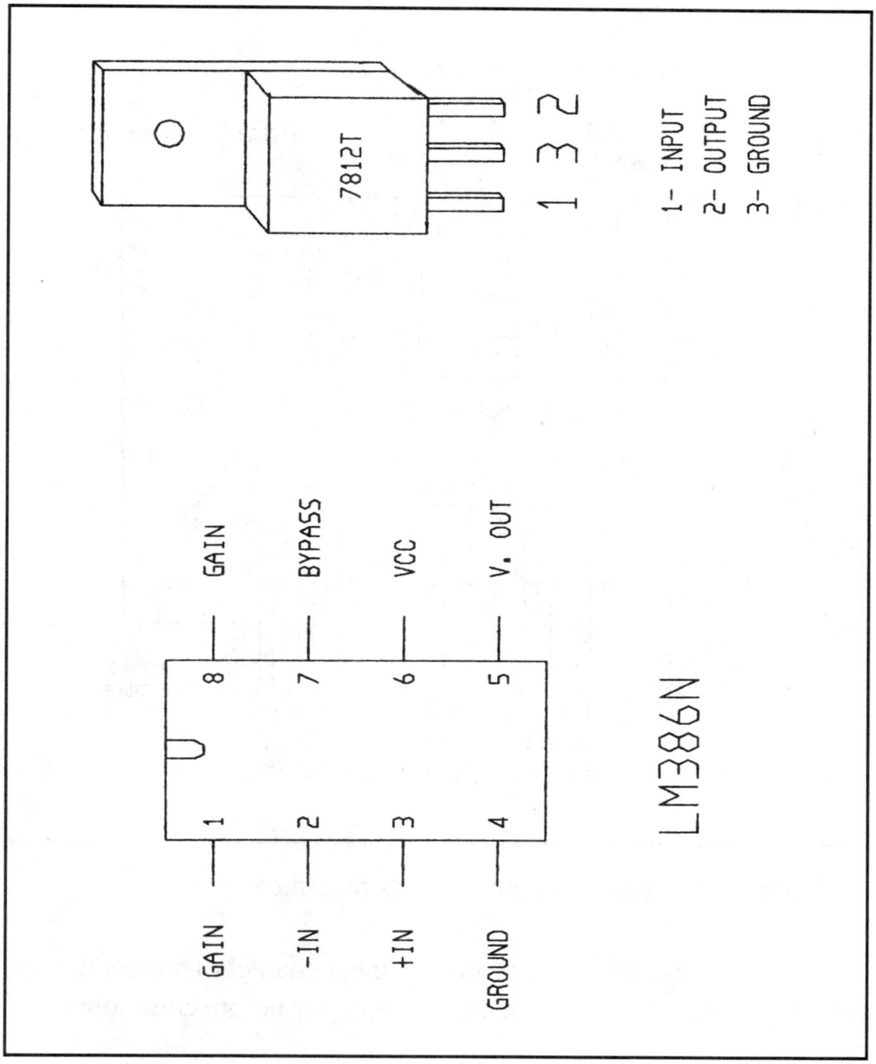

Figure 10-5. LM386 and 7812T pinouts.

I've provided *Figure 10-7* as a template, but you can arrange the panel in any fashion that suites you.

Using drills, a small hobby saw and files, make the various panel openings for the display, switches, LED, banana jacks and potentiometers. Once this is completed, spray paint the panel a light color. When it's dry, label the panel with dry-transfer lettering. Once you're finished

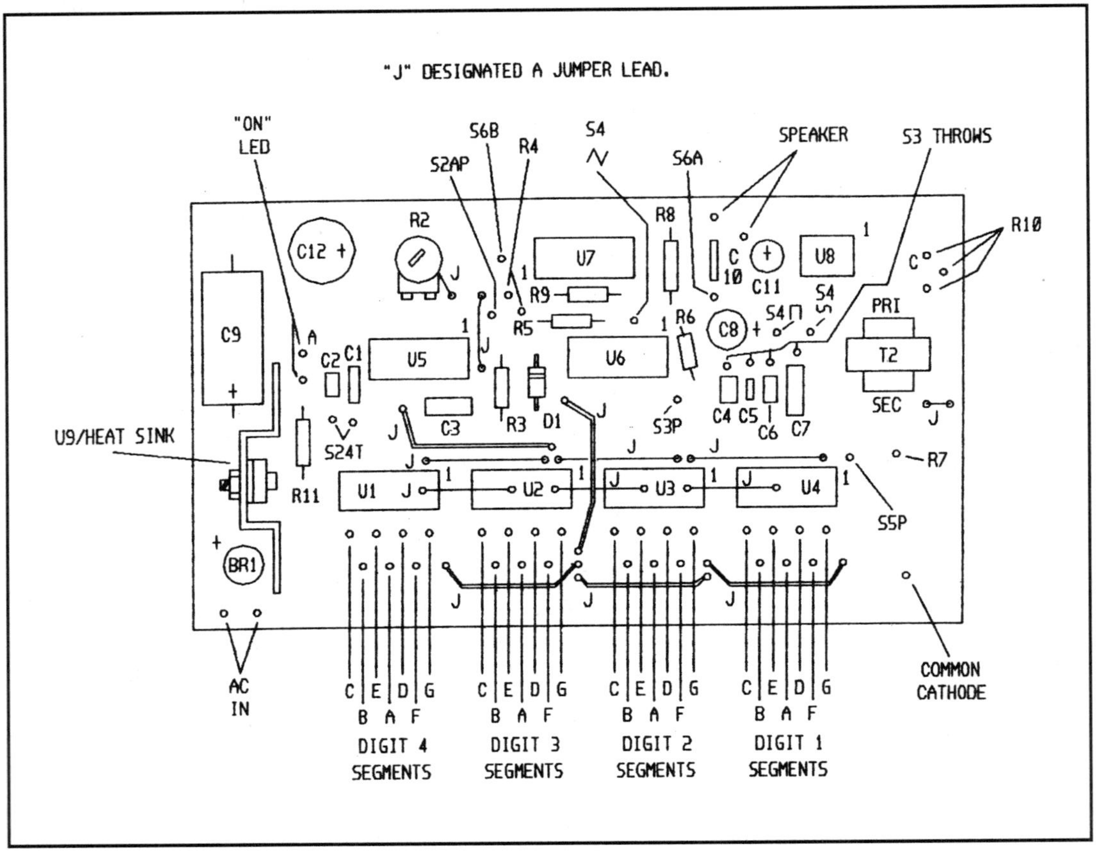

Figure 10-6. Function generator parts placement.

with the labelling, apply a light coat of clear enamel to protect the delicate dry transfer. Let this coat cure completely before continuing.

As for the case *shell*, very little needs to be done. The bottom section needs holes to secure the power transformer (T1), and you should drill a circular pattern of small holes in the top half to accommodate the speaker. If you plan to paint the shell, now is the time. For the prototype, I used a medium blue for the case and yellow for the front panel. This is, of course, a matter of personal preference.

The back panel needs only a 3/8" hole for a rubber grommet. Run the power cord through the center of the grommet and knot it on the inside to furnish strain relief. Now connect the cord to the power transformer

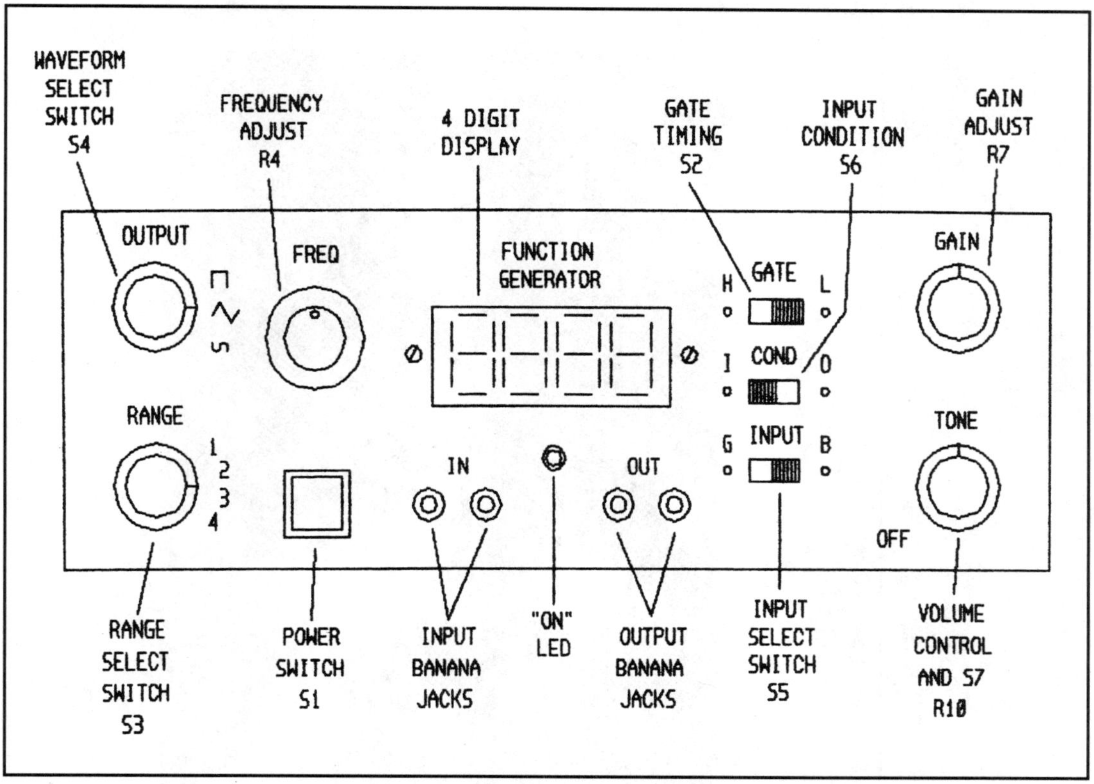

Figure 10-7. Front panel template.

primary leads and use an appropriate length of ZIP cord for the panel-mounted power switch.

When the case is ready, secure the panel components and display in place, and solder the banana jacks and power switch to their respective leads. Attach the main board with screws to the mounting posts, and slide the front and back panels into the slots in the shell bottom. Also, using epoxy, cement the speaker to the inside of the top section so that it covers the previously-drilled holes.

Hold off on putting the top in place since the unit still has to be calibrated. Other than this, and connecting the speaker, construction is complete (*Photo 10-1*).

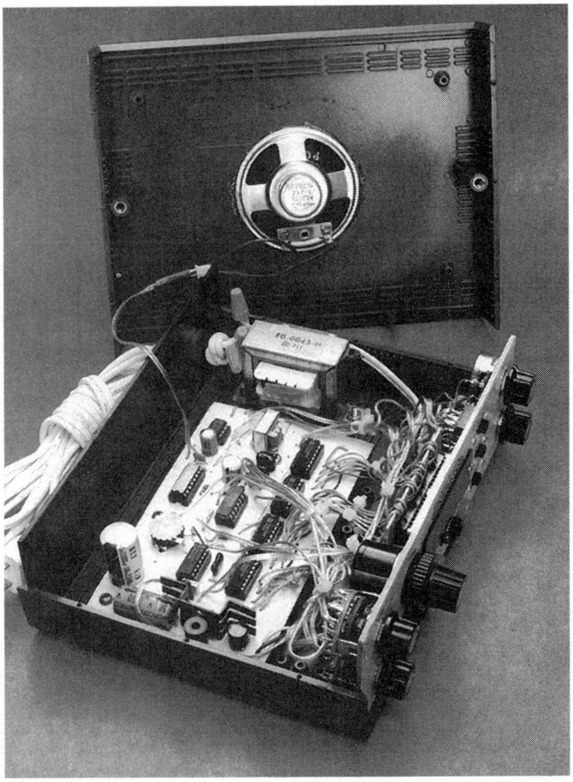

Photo 10-1. Internal view of the digital function generator. Note the speaker on the top half of the case.

Testing and Application

Plug in the unit and apply power. The LED should light up and the display should be showing something besides *0000*. If you're getting four zeros, make sure the function output switch (S4) is set to the square or triangular wave position. Some of the other settings will also affect the display, such as the gate timing capacitors controlled by switch S2, or the range capacitors connected to switch S3. Concerning the latter, for the prototype, I labelled the four settings *1*, *2*, *3* and *4*, with *4* addressing the smallest value capacitor. Remember that the smaller the

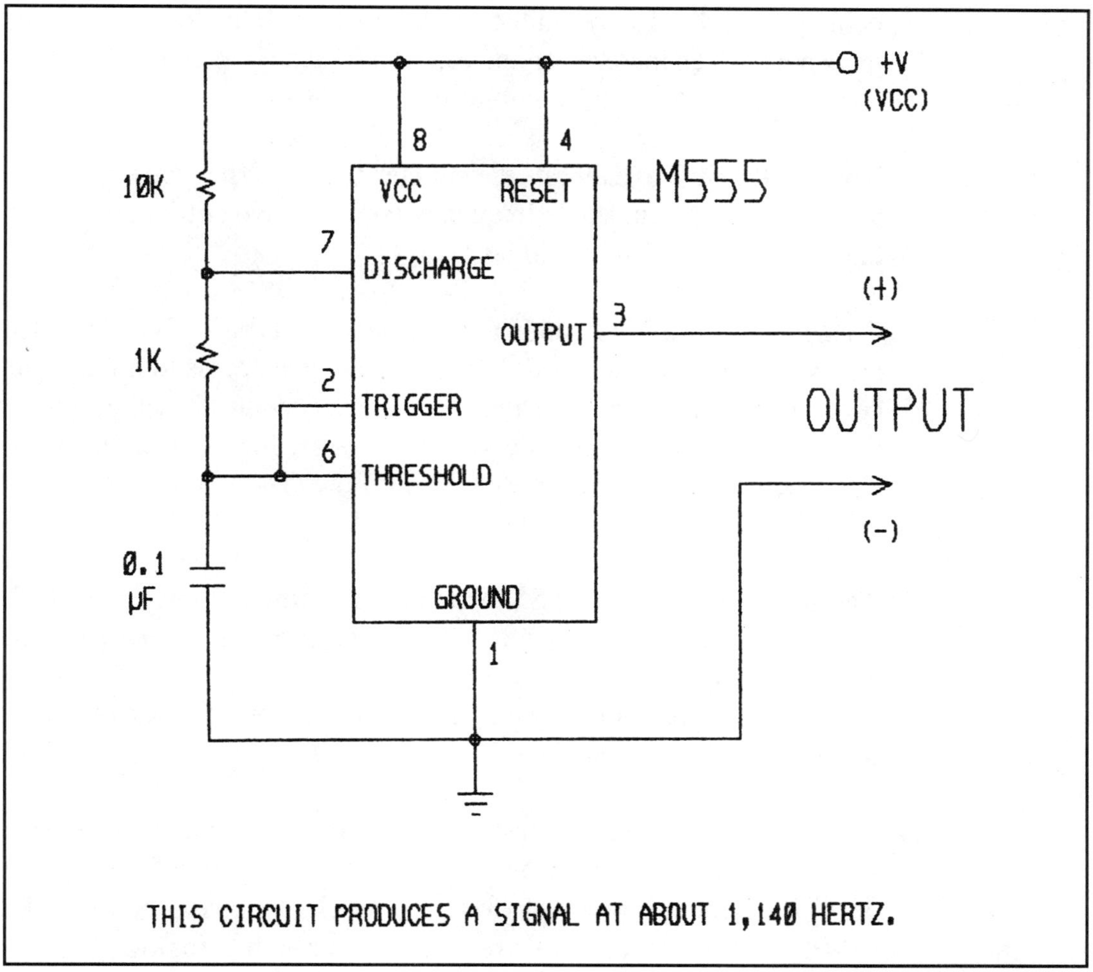

Figure 10-8. LM555 frequency reference.

capacitor, the faster the charge/discharge cycle; so the smaller the capacitor, the higher the frequency generated.

Gate timing works on the same principle, only sort of in reverse. The larger the capacitor value, the higher the resolution. This is due to a longer counting period.

Next, set the range to *2* and turn on the audio amp with switch S7. If all is well, you will hear an audio tone representative of the frequency being produced. As you adjust potentiometer R4, the tone will change in response to an increase or decrease in frequency. When you're done, rotate R10 to the *off* position to disconnect U8's power.

Now it's the time to calibrate the counter circuit. What you will need is a source of a known, stable frequency (square wave or triangular in nature), preferably in the audio range.

Figure 10-8 shows an LM555 arrangement that is easily breadboarded, and will provide about 1,140 hertz. If you have access to a frequency counter, confirm the 555's actual frequency, then introduce the signal to the *in* banana jacks. Be sure switch S5 is set for external input (B). Next, adjust PCB potentiometer R2 until the display reads as close to 1,140 hertz as possible. And that is that.

Another approach is to set S5 to the internal setting, then read the 8038 output with a frequency counter. Now you can adjust potentiometer R2 to match the reading. Either method works quite well, and drives home the aforementioned point regarding the value of more than one frequency counter.

Put the lid on the mighty beast, and it's ready for service. Since panel space was at a premium, I kept the labelling to a minimum on the prototype. Everything is pretty obvious, but for clarity, I used the following with the slide switches: *Gate* for timing; *H* for high; *L* for low; *Cond* for input conditioning; *I* for in; *O* for out; *Input* for the 4026 clock; *G* for generator (the 8038); and *B* for bypass (the external input).

The applications should be fairly obvious by now: tone generation, frequency standard, signal injection, oscilloscope trace patterns, just to name a few. Additionally, the on-board counter will be very helpful when medium-accuracy readings are acceptable. The gate times of 1/10 second (C1) and 1/100 second (C2) are fast enough to enable you to use the counter for audio frequency readings. Even with a rapidly-changing signal, such as music, this counter should be very useful.

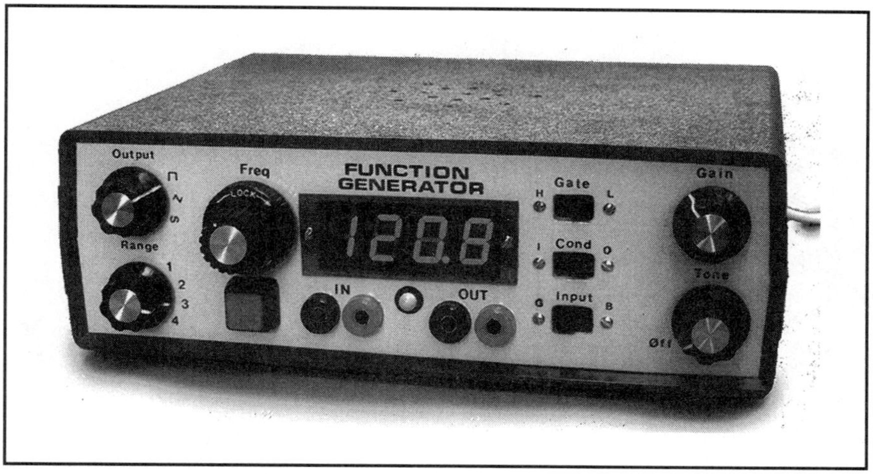

Photo 10-2. The completed digital function generator.

Another highly useful application will be checking the frequency response of amplifiers and speakers. If you're into audio, this generator will be a tool of endless value. After using it a time or two, I'll lay odds you'll give it very high marks.

Conclusion

Chalk one up for the ol' test bench: a digital function generator at about one quarter the cost of similar commercial units (*Photo 10-2*). Best of all, you can configure this device with the whistles and bells YOU want, instead of what the manufacturer wants.

Now, I'm well aware that function generators aren't a number-one priority for many hobbyists; but once you have one on your bench, you will quickly learn to appreciate the device. The convenience of variable frequency sine, triangle and square waves, at your beck and call, greatly enhances your lab's versatility. Just think of all the neat stuff you can do with this thing. You can generate the three most common waveforms, read frequencies and test audio equipment. What more could you want? If nothing else, this unit makes a great sound-effects box for your home videos.

Chapter 11
An Eight-Digit Frequency Counter

Introduction

Few items on a test bench are more useful than a frequency counter. These counters take an input signal, determine its frequency, then employ either analog meters or digital readouts to display the result. Both approaches work, but those pretty red numbers in a digital readout are hard to beat for legibility, so that's the route our project will take.

Not too many years ago, frequency counters were monumental task machines costing thousands of dollars. Each counting stage, as well as the support circuits, had to be individually constructed from transistors or (before that) tubes. Digital displaying was still the stuff of Buck Rogers.

Today, however, through the auspices of integrated circuits (ICs) and other microelectronics, the venture is far less complex and comparatively economical. Our counter will use four ICs, a handful of discrete components, and ten switches for a compact, functional design. The construction cost is well under a hundred dollars, and the unit can be assembled in a few evenings. Best of all, once you have the counter operational, I think you, too, will learn to love the little beast.

Theory

The term *frequency counter* applies to any event counter that is governed by a very precise timing period. This period is referred to as the

FREQUENCY COUNTER PARTS LIST

SEMICONDUCTORS

U1	74HC4049 Hex Inverter
U2	74LS90 Decade Counter/Divider
U3	ICM7216DIPI 8 digit Frequency Counter (Harris/Intersil: Digi-Key, Mouser, DC Elect.)
U4	7805T Positive 5 Volt Regulator
LED1	Green T-1 3/4 Jumbo Light Emitting Diode
LED2	Red T-1 Light Emitting Diode
LED3	Green T-1 Light Emitting Diode
DSPL	8 Digit Common Cathode LED Multiplexed Display (Can be 4-MAN6740's multiplexed)
D1	1N4733 5.1 Volt Zener Diode
D2-5	1N914/1N4148 Silicon Signal Diodes
BR1	200 Volt/1 Amp Bridge Rectifier (W02G, or Equiv.)
XTAL1	10 Megahertz Quartz Crystal

RESISTORS (All Resistors 5% Carbon Film)

R1	47,000 Ohm, 1/4 Watt Resistor
R2	100,000 Ohm, 1/4 Watt Resistor
R3	10,000,000 Ohm, 1/4 Watt Resistor
R4	3,000 Ohm, 1/4 Watt Resistor
R5	220 Ohm, 1/4 Watt Resistor
R6, 7	10,000 Ohm, 1/4 Watt Resistor
R8	20,000,000 Ohm, 1/4 Watt Resistor (2-10M in Series)

CAPACITORS

C1, 2	0.22 Microfarad Mylar Capacitors
C3	2,200 Microfarad Electrolytic Capacitor
C4	2.2 Microfarad Electrolytic Capacitor
C5	100 Picofarad Disk Capacitor
C6	0.1 Microfarad Mylar Capacitor
C7	39 Picofarad Disk Capacitor
C8	5-35 Picofarad PCB Trimmer Capacitor

SWITCHES

S1	SPST Slide Switch
S2	SPDT Slide Switch
S3	DPDT Slide Switch
S4, 8	Normally Open, Momentary Contact Pushbutton
S5, 6, 7	DPDT Push On-Push Off Switches
S9	2P6T Rotary Switch (Only 4 T's Used)
S10	SPST Push On-Push Off Switch

OTHER COMPONENTS

T1	120 VAC Primary/ 12 VAC Secondary Power Transformer
J1, 2	RCA Jacks
F1	1 Amp Fuse
CASE	7.9" x 6.3" x 2.5" Instrument Style Case (H2505, or Eqv)

MISCELLANEOUS

IC Sockets, LED Holder, Solder, Hook-Up Wire, 120 VAC Line Cord/Plug, PCB Materials, Hardware, Paint, Dry Transfer Lettering, Clear Red Plastic, Bumper Feet, Knobs, Fuse Holder, Etc.

timing *window* or *gate*, and the more precise it is, the more accurate the count. For this reason, any good frequency counter is crystal-controlled.

The most common timing window is one second, which results in a reading of the input in hertz. Let's say you introduce a 453-hertz signal to the counter. This means the signal will cycle 453 times in one second, giving a display of *453* during a one-second timing gate/window.

How does the device determine that 453 cycles have occurred? The answer lies in the *decade* counter/divider. These devices will count from 0 to 9, then reset. In our system, there will be eight of them cascaded to produce the eight-digit resolution. This counting action, in effect, results in a division by 10. As we go along, this process will become much more clear.

The first counter divides 453 by 10, with the result of 45 and a remainder of 3. The 45 is passed on to the next counter, and the remainder is displayed as a *3*. This outcome is due to the fact that the decade counter can only count evenly up to "10" 45 times.

Stage two now divides 45 by 10, for the result of 4 and remainder of 5. Again, the display gets the 5, and 4 is passed on to the third counter. Since 4 cannot be divided by 10, it becomes the remainder, and is sent to

To save space and reduce pin numbers to a minimum, time multiplexing is employed for the range, control switches and decimal point location. This allows the digit driver to handle several functions by means of special timing. In this fashion, the digit pin only recognizes a command during a specific section of the timing cycle.

GATE TIMING:

Pin 14 Connected To	Window (Seconds)	Number of Cycles
Pin 3 - Digit 1	0.01 Sec.	One
Pin 5 - Digit 2	0.1 Sec.	Ten
Pin 4 - Digit 3	1 Sec.	One Hundred
Pin 6 - Digit 4	10 Secs.	One Thousand

CONTROL INPUT:

Pin 1 (Control In) Connected To	Function
Pin 11 - Digit 8	Tests All Display Segments
Pin 4 - Digit 3	External Decimal Point Enable
Pin 3 - Digit 1	External Oscillator Enable
Pin 6 - Digit 4 and Hold	Blanks the Display

Table 11-1. 7216 operational specifications.

the display as the third digit. Thus, the reading will now be *453*, or a compilation of all three remainders.

Naturally, the process continues for as many digits as are needed to display the full input signal. If the input is 27,862 hertz, the system utilizes five of the eight counter/divider stages, while 8,759,640 hertz will require seven stages.

The next consideration is the actual length of the timing window. With our design, four such windows are available, and each affects the accuracy or resolution of the displayed frequency. Each also controls the speed of the measurement.

So, as with most electronic gadgets, a compromise is necessary. The four timing gates are 0.01 second, 0.1 second, 1 second and 10 seconds, with display resolutions of 100 hertz, 10 hertz, 1 hertz and 0.1 hertz, respectively (see *Table 11-1*). As can be seen, the faster the measurement, the less resolution. However, this is not really a problem in most cases, as a one-second gate is quick and provides accuracy within 1 hertz. So, for many tests, you will probably end up using the one-second window.

Once the counts are performed, their results have to be displayed. As stated earlier, this can be done on an analog meter movement or by seven-segment, light-emitting diode (LED)/liquid crystal display (LCD) arrays. For the sake of accuracy and convenience, most commercial counters use the latter; and if it's good enough for them, it's good enough for us. Actually, with present-day ICs, it would be more difficult to do it with a meter. Not impossible, but definitely more complicated. This is a quasi-repercussion of the age: the digital nature of most counter ICs, especially devices dedicated to frequency counting.

Speaking of which, let's meet the *brains* of our project: the Harris/Intersil ICM7216DIPI eight-digit frequency counter. This is a true marvel of miniature electronics. On a single chip, there is a decade time-base counter, an eight-decade data counter and latches, seven-segment decoders, digit multiplexers, eight-digit drivers, and a high-

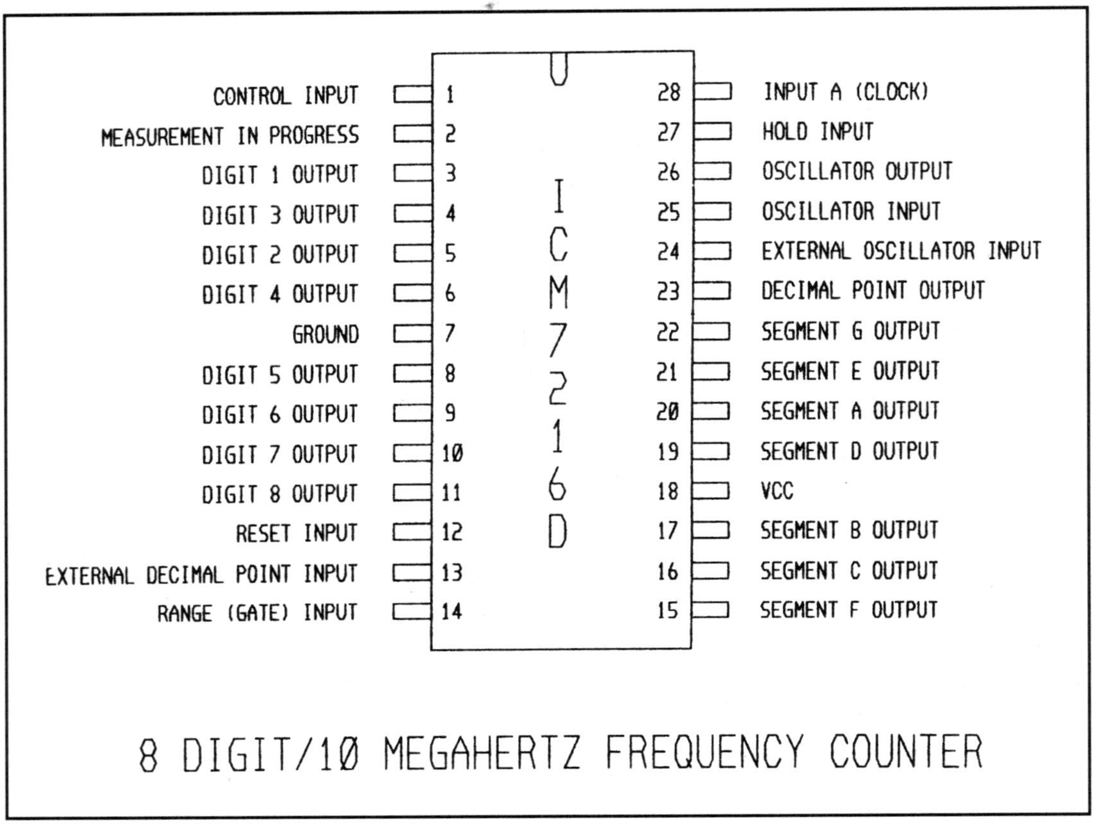

Figure 11-1. ICM7216DIPI pinout diagram.

frequency oscillator. Additionally, *overflow* and *measurement in progress* signals can activate panel-mounted LEDs.

The maximum input frequency is 10 megahertz. The IC provides four range selections (gate times), as well as automatic decimal point positioning for each setting. The digit drivers can handle common-cathode LED arrays up to instrument size (0.56”). All readings are displayed in kilohertz.

Through internal timing, the multiplexers control not only the display but decimal point placement, control functions and range. The internal timing feature allows the entire circuit to be housed in a standard 28-pin dual in-line pin (DIP) package (*Figure 11-1* shows the pinout).

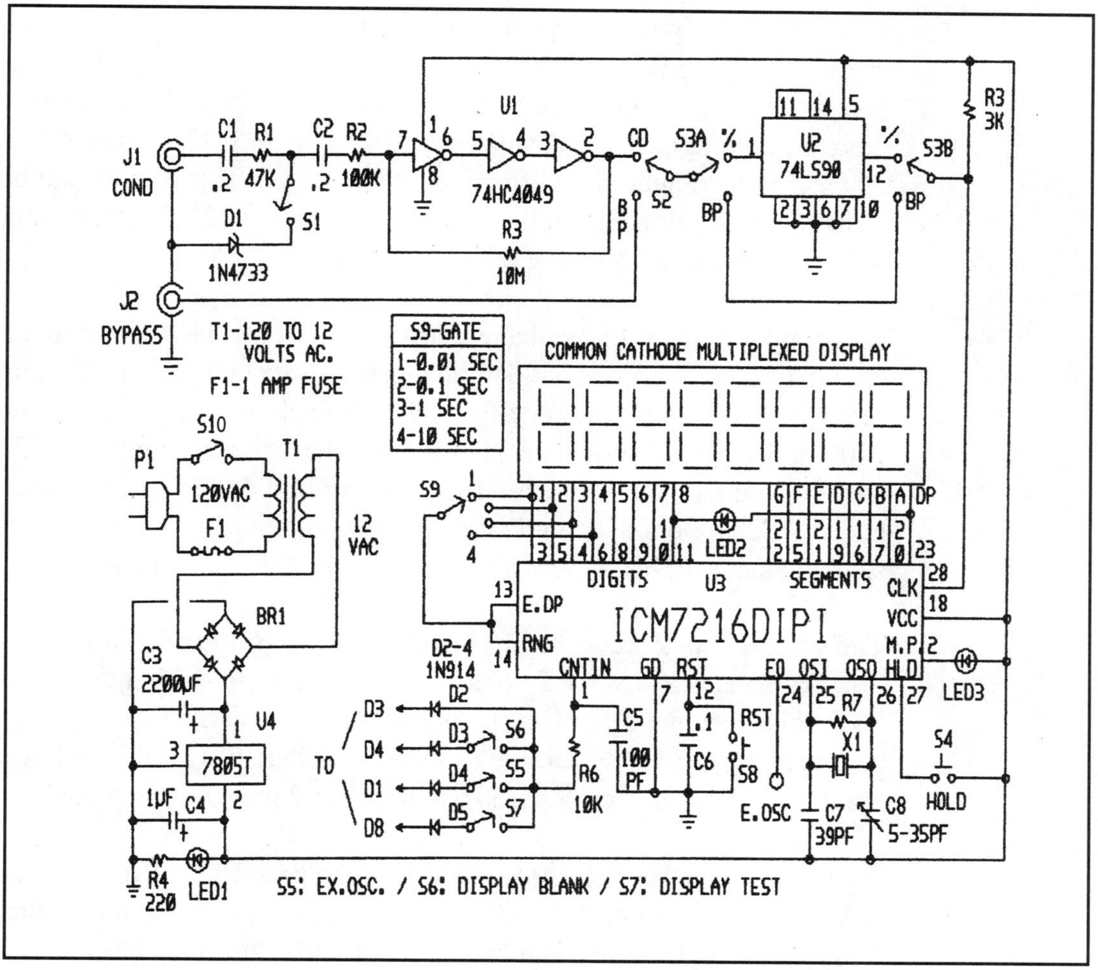

Figure 11-2. ICM7216D frequency counter schematic.

It's easy to see why I call this chip a *marvel.* With very little external support, you can construct a highly accurate eight-digit frequency counter. What a deal!

In keeping with the previous discussion about counter operation, the 7216 takes the input signal and successively divides it through as many of the eight stages as necessary. Employing the decoders, multiplexers and drivers, the division remainders are conveniently displayed on an LED array. In short, the frequency of whatever signal you input is going to appear on the display, up to 10 megahertz.

In reality, the chip is guaranteed to 10 megahertz; but in the past, as long as the signal was a clean square wave, I have been able to measure inputs as high as 15 megahertz. The 10-megahertz limit does present a drawback, though, so the design (as seen in *Figure 11-2*) implements a *divide-by-10* counter circuit to increase the range. This divider can be switched in or out between the signal input and the 7216's clock input (pin 28), effectively dividing the input by 10.

There are a number of available chips that perform this service, but a couple of considerations have to be taken into account when choosing the IC. First is the frequency limitation of each candidate. Since the 7400 transistor-to-transistor logic (TTL) series chips function to up 20 megahertz guaranteed (and usually a little beyond that), they are a good choice. At the opposite end is the 4000 complementary metal oxide semiconductor (CMOS) series that cuts off at 10 to 15 megahertz.

The second consideration when choosing an IC is operational voltage. The 7216 wants a positive 5 volts, which fits perfectly with the 4.75-to-5.25 voltage range of TTL semiconductors. The CMOS devices will also operate at 5 volts; but due to their characteristics, at this voltage, maximum frequency response is down to 1 or 2 megahertz at best.

So, to be at all effective, the divider needs to be a TTL. I selected the 74LS90 for power conservation, so we now have a maximum counter frequency of between 20 and 25 megahertz. To carry this one step further, a 74HC160 could take this up to 40 or 45 megahertz, but will need a different configuration.

Please note: if you can still find one, the 11C90 prescaler counter/divider will handle frequencies in the hundreds of megahertz. These are nice, but be prepared to pay almost as much for the prescaler as for the 7216.

One drawback to digital counters is that they only recognize a digital or square wave input; triangular (sawtooth) and sine waves will not register or be counted. The way around this obstacle is to *precondition* the input signal. As you may remember from the last chapter, preconditioning masquerades other waveforms to appear as square waves.

Again, many frequency counters employ op-amp circuits for preconditioning; but these are often inefficient and usually require a split (dual-polarity) power supply. All of this makes them difficult to work with. So, once again we will utilize Mr. Bolles' excellent 74HC4049 design.

As before, gain is kept high enough so that the input signal saturates the chip, thus producing fast rise and fall times. Additionally, running the 74HC4049 rail-to-rail, so to speak, *clips* the waveform tops to simulate a square wave, at least well enough to fool the 7216.

Now the counter will accept and read any type of input signal. Of course, the same input voltage restrictions apply as before. If you plan to introduce a voltage higher than +5, switch on zener diode D1 for regulation.

Speaking of switches, let's take a look at them. As seen in *Figure 11-1*, S1 is for the zener diode, and S2 determines whether or not the input is conditioned. S3 takes S2's output and sends it straight through or to the 74LS90 divider.

S4, a normally open pushbutton, handles the display *hold* function, while S5, 6 and 7 control *external oscillator input*, *display blanking* and *display test*, respectively. S8 resets the counter for a new measurement, and S9 selects the gate timing. S10 is the power switch.

I'm the first to admit there are a lot of switches, but in their defense, they add a lot of versatility. From the front panel, you can supervise all functional aspects of this counter; a feature you will appreciate from the outset.

Regarding power, the 7216 requires only a positive 5 volts, as do the 74LS90 and 74HC4049. However, I'm not always sure this characteristic is appropriately cherished. I remember some of the early chips that wanted +5V, +12V, -12V and so forth. I also remember the associated *nightmare* of a power supply. So, for what it's worth, the single voltage approach really makes life simpler.

With our circuit, power is derived from a standard linear-regulated supply. The 7805T regulates the 12-volt product of transformer T1 and rectifier BR1 into a nice, clean positive 5 volts; and of course, this remains stable. The last point is very important for any type of test equipment.

There you have it: the operation of the ICM7216 frequency counter. Through the magic of integrated circuits, all the necessary components of a once-bulky and complex tool can be housed in an 8" x 6-1/2" x 2-1/2" case. Next, we will talk about how to put all this neat stuff in the case.

Construction

As with most of the projects in this book, this one can be laid out on a printed circuit board (PCB), or with point-to-point wiring on perf board. I have always preferred to use PCB, unless there is a good reason not to; so this is the approach I chose for this device. *Figure 11-3* illustrates the foil (solder) side of the PCB, while *Figure 11-4* is a parts placement guide.

With that said, let's get to work. First install the IC sockets and jumpers, then install the discrete components and regulator U4. I saved the diodes and crystal for last as they are either heat or vibration sensitive, and thus susceptible to damage.

With everything in place, set the main board (*Photo 11-1*) aside for the moment and turn your attention to the case. Preparing the case at this point allows for the installation of the panel components and PCB before you have to connect the many hookup leads. In this fashion, each wire can be cut to the exact required length, which leads to a cleaner, less costly finished product.

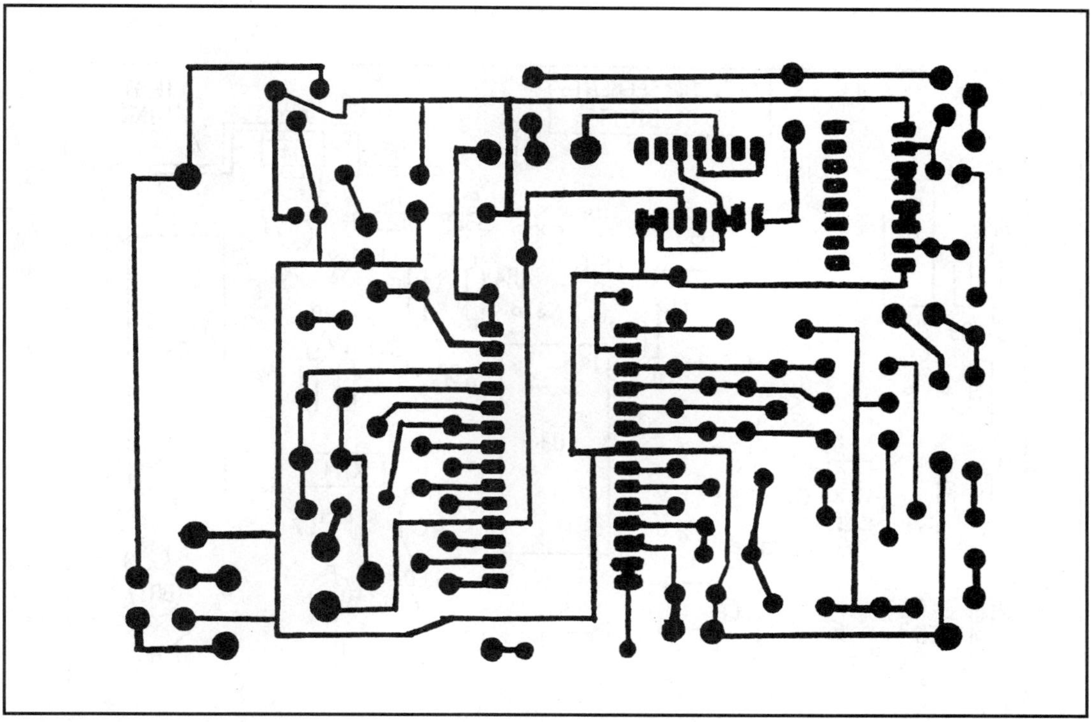

Figure 11-3. Frequency counter foil pattern.

A quick note here: when I say *exact required length*, I'm not necessarily talking about a straight line from point A to point B. You will want to leave some slack so the wire bundles can be properly routed throughout the case.

Using the template in *Figure 11-5*, drill or cut the holes needed to accommodate the various panel devices. For the long, rectangular display opening, a Dremel-style tool equipped with a saw blade works well. Drill small holes in each corner, then your saw should easily cut the plastic in between the holes. Hobby files complete the aperture by smoothing and shaping it to match the display.

For the smaller, rectangular cutouts for the slide and pushbutton switches, start with round holes and shape them with the files. Naturally, all of the round holes can be done with standard drill bits.

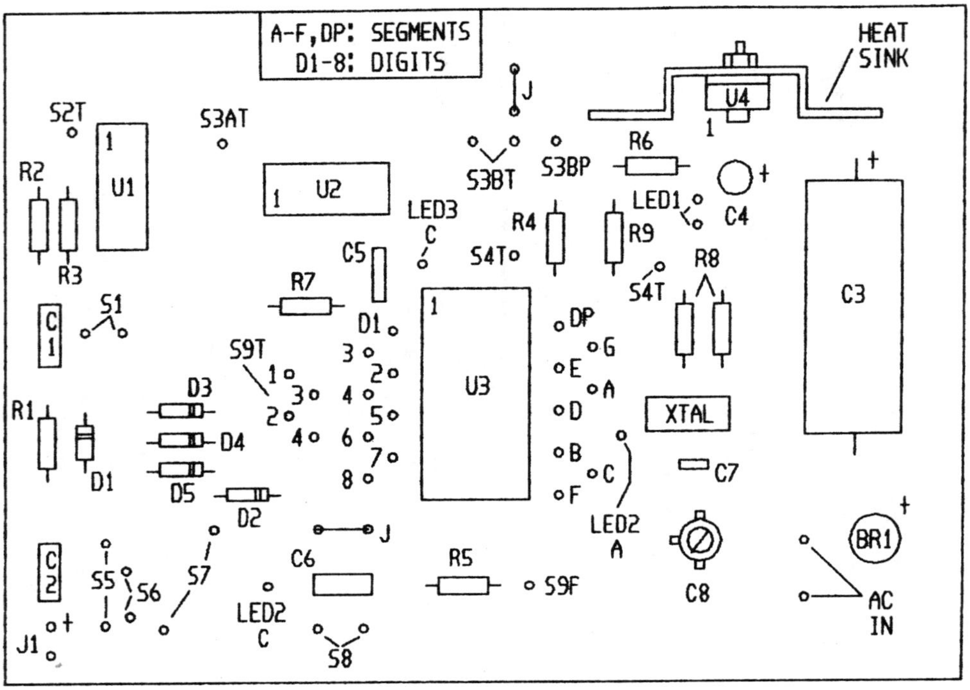

LED1- "ON" INDICATOR
LED2- OVERFLOW
LED3- MEASUREMENT IN PROGRESS
S1- HI/LO VOLTAGE
S2- INPUT COND/BYPASS

S3- DIVIDE BY 10
S4- DISPLAY HOLD
S5- EXTERNAL OSCILLATOR
S6- DISPLAY BLANKING
S7- DISPLAY TEST

Figure 11-4. Frequency counter parts placement.

Once you have the front panel ready, paint it and apply the dry-transfer lettering once the paint is cured. Then spray on a light coat clear enamel to help protect the labelling.

Next, drill and paint the *shell* section of the case, and drill the back panel. You can leave the back panel as its natural color, since it merely houses the power cord access and fuse holder. The shell requires only two holes in the bottom half to secure the power transformer.

When all of the paint is completely dry, install the panel components, bolt down the transformer and the fuse holder, and position the AC cord. Put the PCB in place, but leave it loose for the moment, as you still need to access the solder side.

Now come the hookups. Start by determining the lead lengths for the input jacks and switches. Once cut and stripped, the leads can be soldered to the board and the appropriate panel device. It is also advisable to use a variety of insulation colors for easier lead identification, and to secure the wire bundles with small cable ties.

Continue to connect everything until only the display is left. As for the display, it is an eight-digit, multiplexed, common-cathode LED array of any size up to and including 0.56." There are several ways to go here. For the prototype, I used an eight-digit multiplexed display

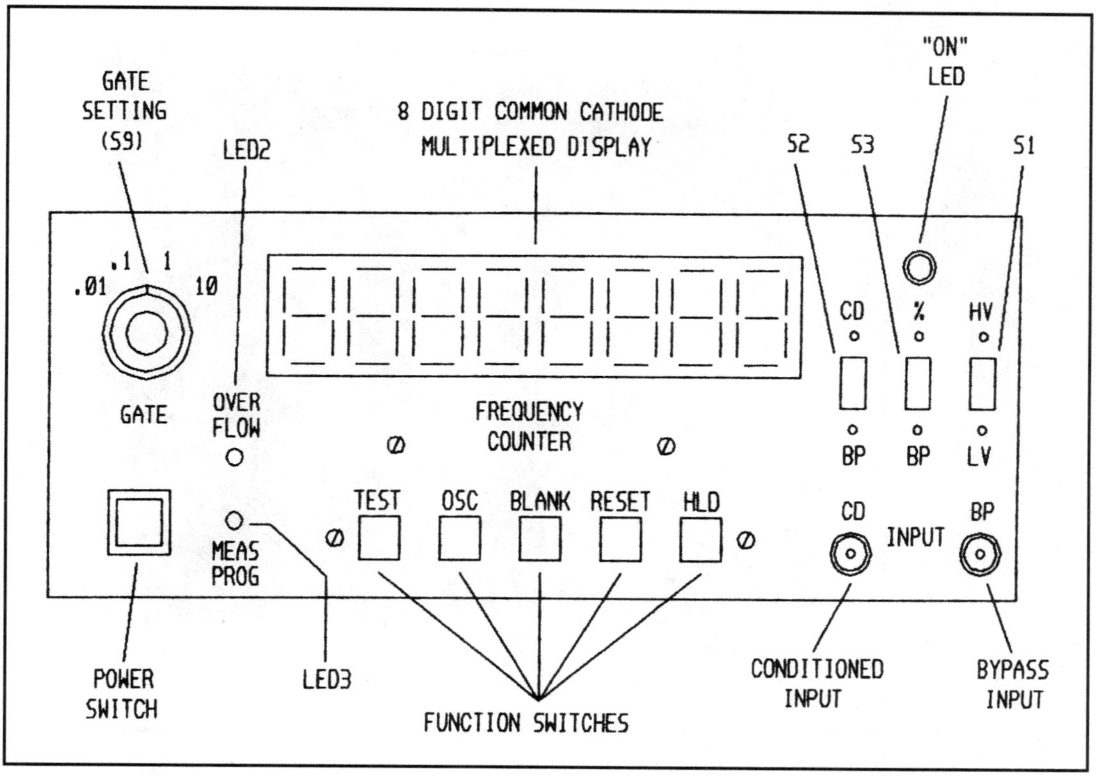

Figure 11-5. Frequency counter panel template.

robbed from an old cable TV control device. You can also wire eight separate displays or four dual displays into a multiplexed arrangement. In this context, *multiplexing* means that all of the *a* segments, *b* segments, *c* segments, etc., are wired together to form eight-*segment* lines, while the individual common cathode terminals comprise the eight-*digit* lines.

Another possible source for a display is the surplus market. At one time, several companies made four-digit multiplexed display arrays, and you might be able to find some of these. A problem here is that in recent years, LED displays have taken a back seat to LCD displays, and are not as plentiful as they once were.

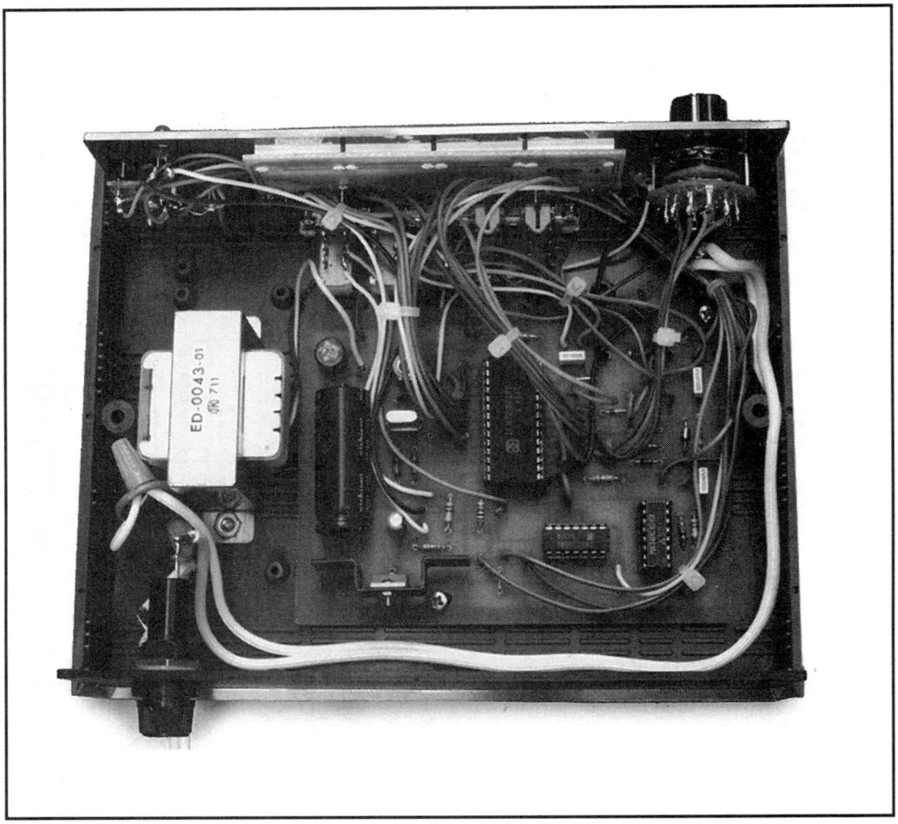

Photo 11-1. Internal shot of the completed frequency counter.

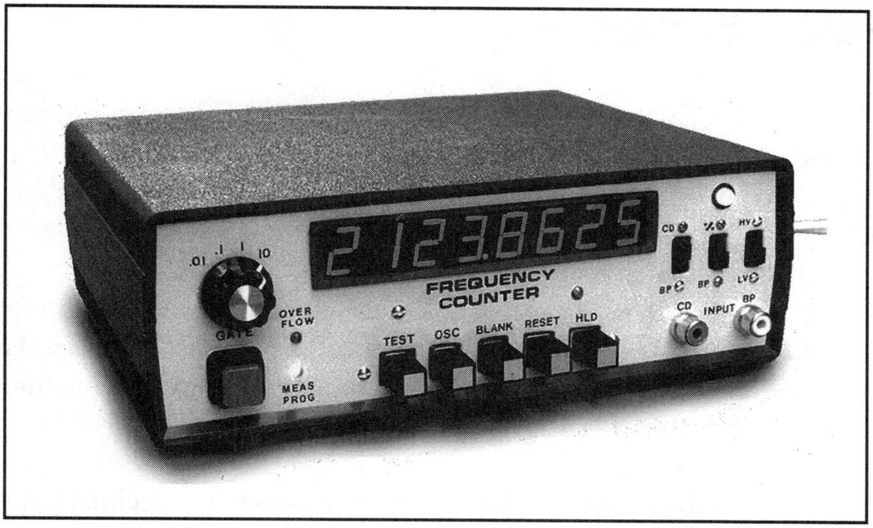

Photo 11-2. External view of the completed frequency counter.

However, the worst case scenario for the display is having to wire eight separate seven-segment displays, which is tedious but not an overwhelming task.

Anyway, with the display complete and in place, connect precise lead lengths to both the PCB, and the digit and segment terminals. This step should complete the installation of the counter, but don't forget to connect the transformer leads. Finally (I know this sounds silly), put a fuse in the holder. There's an old story about a guy who called a TV repair tech to his home (back in the good ol' days when they would come to your home) because his set didn't work. After arriving, the service individual quickly discovered the problem: the TV was unplugged. Laugh if you must, but several repair persons have told me it's a true story. In fact, it happens more often than you might expect. The moral: power is relatively important.

Photo 11-1 shows the prototype at this stage. We're now ready to test and calibrate the counter, so don't button up the case just yet. Both procedures are quite simple, so let's get to it.

Testing and Calibration

With the ICs out of their sockets, plug in the counter, hit the power switch, and the *power on* indicator should light up. If not, be sure the LED is properly connected, then check for +5 volts at the power supply's output.

Next, examine each IC socket for positive voltage at pins 1 for U1, 5 for U2, and 18 for U3. If all is in order, turn off the power and install the integrated circuits. As always, be sure to observe proper orientation, particularly with the 7216.

Return power to the circuit. The display should show a decimal point (DP) followed by zeros. The location of the decimal point depends on the setting of switch S9. At the 0.01 setting, the DP will be at the second position accompanied by a single zero. At 0.1, the DP moves to the third position with two zeros, and so forth.

As you change S9, watch the *measurement in progress* LED. This indicator will blink briefly, as there is a 200-millisecond delay between each measurement. During this time, LED3 is off. Also notice that as the gate time decreases, the diode blinks more rapidly. This, of course, is due to the increased frequency of the measurements.

Now, activate *test* switch S7, and the display will fill with decimal points and eights. The *overflow* indicator, LED2, will also light up. This tells you (among other things) that all display segments are functioning and all driver circuits are operational.

With that done, we now have to introduce a signal of known frequency to the counter. If you don't have a source readily at hand, *Figure 11-6* illustrates a simple crystal-controlled inverter clock circuit that does nicely. This oscillator can be breadboarded in a matter of minutes, and will give you a highly stable signal.

Set the gate to 1 second, and input the clock signal through bypass jack J2. Be sure that both S2 and S3 are set at BP (bypass). Hit the reset

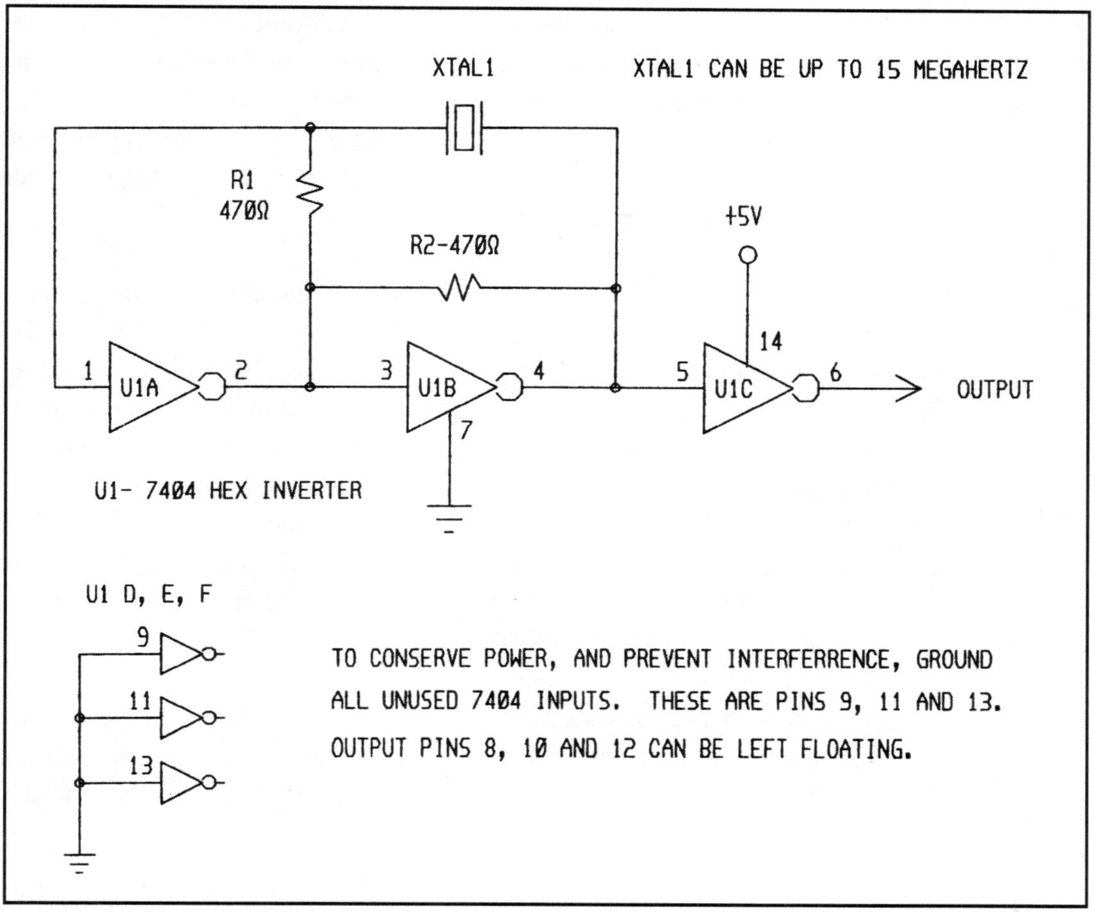

Figure 11-6. 7404 inverter clock circuit.

button, S8, and at the end of the 1-second timing cycle, the display will read the frequency of the crystal. Watch LED3 for the blink signaling the end of the measurement.

Don't become too concerned, however, if the reading is a little low. Most crystals are guaranteed to a tolerance of 0.05 %. It has been my experience that they usually oscillate slightly below their actual marked frequency. For example, a 4-megahertz crystal might give a display of 3.9754 megahertz, but this is well within the tolerance of 3.8000 to 4.2000 megahertz.

The next step is to adjust trimmer capacitor C8 until the display reads as close to the advertised frequency as possible. Use a nonconductive tool and make gradual adjustments. You may want to go to a 0.1-second gate (at least at first), then switch back to a higher window for the *fine tuning*. Again, you'll probably never get a 4-megahertz crystal to produce an exact 4-megahertz reading.

As each reading comes up on the display, you may notice changes in the least significant (first) digit. This is normal, and is influenced by ambient temperature changes and oscillator stability. However, activating *hold* switch S4 will freeze the display to the latest measurement until the hold is released. This can be useful when measuring erratic or rapidly-changing signals. Additionally, by triggering the *blank* switch (S6) then *hold*, the display will blank out. All measurements will halt until the switches are disengaged. You will note that any time the hold or reset switch is activated, the *measurement in progress* LED extinguishes, indicating measurement termination.

If everything is as it should be, close up the case (*Photo 11-2*). Your frequency counter is ready for service. At this point, operation of the unit should be fairly obvious; but in the coming section, I will briefly cover the finer points.

Operation

The first step is to plug in the counter. I know, there I go again; but the thing don't work well without power. Now hit the main switch, and you will see the same results as before.

Next, introduce the signal you want to test. If you're not sure about the waveform or voltage, start with the conditioned input (CD) and the high-voltage (HV) switch settings. Once a reading is established, you can change inputs to determine the nature of the signal.

If you know the signal is a square wave at +5 volts or less, then the bypass input (BP) and associated switch assignments may be employed.

Incidentally, it's not necessary to turn off the unit to change inputs or initiate a new signal.

From here you can adjust the gate to whatever speed and resolution you desire. Again, this can be done at any time, but a change will halt the previous measurement and start a new one. The same is true for the reset and hold functions.

As a quick sidebar, let me digress for a moment, and talk about display blanking. One of the primary purposes of this feature is to free up the LED array for other uses. When this chip was first developed, seven-segment LED displays were quite at bit more expensive than they are now. Thus, it wasn't at all unusual to see (for lack of a better term) a composite of instruments sharing a single display.

Another purpose here is to save power. For a *plug-in* unit like ours, saving power isn't that important; but when the device is battery-operated, you can spare the cells by shutting down the power-hungry LEDs. In the *blank* mode, the rest of the circuit goes into a *low power* status, but no measurements are made during that period. In effect, everything is in *standby* mode.

On a last note regarding power conservation: the display is multiplexed at a rate of 500 hertz. This is to say that only one of the eight displays units is actually on at any given time. However, they are turned on and off approximately 62 (500 divided by 8) times per second, so persistence of vision gives the illusion that all of the displays are on all of the time.

Multiplexing also allows the digit driver lines to handle other jobs. As stated earlier, these include the gate times, control switches and decimal point placement.

To activate the input divider circuit, switch S3 to the upper position. This adds the 74LS90 to the 7216's clock input path. Since the 74LS90 is configured as a *divide-by-ten* counter, any input signal up to about 25 megahertz is divided by 10. Thus, a 25-megahertz signal will appear on the display as 2.5 megahertz. It goes without saying that you must remember to multiply the reading by 10 any time you use the divider feature.

The last point to cover involves the high/low voltage setting of switch S1. In the high *HV* position, zener diode D1 is connected from the input to ground, regulating the input to 5.1 volts. This protects the circuit against damage due to excessive input voltage.

However, zener diodes have a nasty habit of cutting off around 8 megahertz. Alas, signals above this frequency will not register properly, so we are faced with another compromise. In the final analysis, high-voltage signals above 8 megahertz are going to be hard to measure without additional conditioning. The good news is that you probably won't run into very many.

Conclusion

Now that your counter is complete, you're ready for action. Any signal in the range of the unit can be put to the test and have its exact frequency displayed. If nothing else, you will know the signals exceed the range.

At first glance, all this might seem more of a luxury than a necessity, but trust me, you will be surprised by how often this device is needed. For years, I lived without a good frequency counter, but since I've added one to my test bench arsenal, I've often wonder how I existed without it.

Beyond that, the technology you've learned from this project will be invaluable for future endeavors. Counters are a major part of today's electronics, thus a solid comprehension of their attributes is very beneficial. Besides, some handy-dandy prototypes can emerge from the knowledge of these extraordinary digital circuits.

Chapter 12
The Solid-State Oscilloscope

Introduction

The field of electronics took a giant step forward in 1947 when three scientists, John Bardeen, W.H. Brattan and W.B. Shockley, invented the transistor. Solid-state technology had been discovered, and the world would never again be the same. True, before then, some research had been done regarding solid-state diodes, but the transistor opened the door to our present-day electronic accomplishments. Those achievements have led to circuits that use less power, that are faster, more reliable, and of course, less expensive.

From the transistor, the world jumped forward into the realm of integrated circuits, liquid crystal displays (LCDs) and light-emitting diodes (LEDs). I know you are aware of all this, but the point here is that these advances have made things a lot easier for all of us. I offer a prime example: a solid-state, dual-trace (2-channel) oscilloscope. While cathode ray tubes (CRTs) remain the staple of most test bench and laboratory scopes, solid-state displays are definitely becoming major players in the arena.

Most distributors worth their salt carry at least one solid-state display device, which are usually smaller and less power hungry than conventional scopes. These devices are, however, considerably more expensive. I trust that, in time, this will change. But these "new age" oscilloscopes overcome some of the pesky problems that plague their vacuum tube cousins. Probably the most significant improvement is

SOLID STATE OSCILLOSCOPE PARTS LIST

SEMICONDUCTORS

U1, 2	LM3914 Bar Graph Display Drivers
U3, 4, 5	CD4017 Decade Counter/Dividers
U6	NE555 Oscillator/Timer
U7	7812T 12 volt Positive Regulator
LED Display	12 LPT747R 5 x 7 Dot Matrix LED display Arrays with Anode Columns and Cathode Rows (Jameco)
BR1	1-1/2 Amp, 400 Volt Bridge Rectifier (W04M or Equiv)
Q1, 2	2N3904 NPN Signal Transistors
LED1	Green Jumbo T-1 3/4 Light Emitting Diode

RESISTORS

R1, 3	100,000 Ohm Panel Potentiometers
R2, 4, 9,10	1,000 Ohm 1/4 Watt Resistors
R5, 6	10,000 Ohm 1/4 Watt Resistors
R7	4,700 Ohm 1/4 Watt Resistor
R8	1,000,000 Ohm Panel Potentiometer

CAPACITORS

C1	0.01 Microfarad Mylar Capacitor
C2	0.1 Microfarad Mylar Capacitor
C3, 5	1 Microfarad Electrolytic Capacitors
C4	470 Microfarad Electrolytic Capacitor

OTHER COMPONENTS

S1, 2, 4	SPST Slide Switches
S3	DPDT Slide Switch
S5	SP3T Rotary Switch
S6	SPST Pushbutton Switch
T1	120 VAC to 12.6 VAC Power Transformer (>500 mA)

F1	1 Amp Fuse
J1, 2	Banana Jack Pairs

MISCELLANEOUS

PCB Materials, Perf-Board, Solder, Hook-Up Wire, IC Sockets, Hardware, Knobs, Paint, Clear Red Sheet Plastic, Dry Transfer Lettering, Fuse Holder, LED Holder, Heat Sink, Line Cord and Plug, Instrument Case, Etc.

the lack of high voltage for the CRT. Using LCD and LED technology, there is no need for voltages above the normal operating levels.

So, with this in mind, let's build a scope that employs LED matrix displays. These are easily acquired, reasonably priced and very simple to use. While the final resolution will not be quite as good as a CRT display, you will be amazed at the results. Besides, I wouldn't even begin to know where to point you regarding cathode ray tubes and their associated components.

Theory

Let's start by looking at the basic components of the average oscilloscope. In order to display a desired waveform, the system must have both vertical and horizontal scanning, much like your television. Additionally, you will want to be able to control the scanning in a fashion that allows for precise measurement. This involves vertical gain and horizontal frequency (sweep rate).

A third factor to consider is synchronizing the vertical and horizontal functions. In order to keep a consistent display, both elements of the scan must start at the proper place and time. Again, this equates to the requisite timing involved in video equipment.

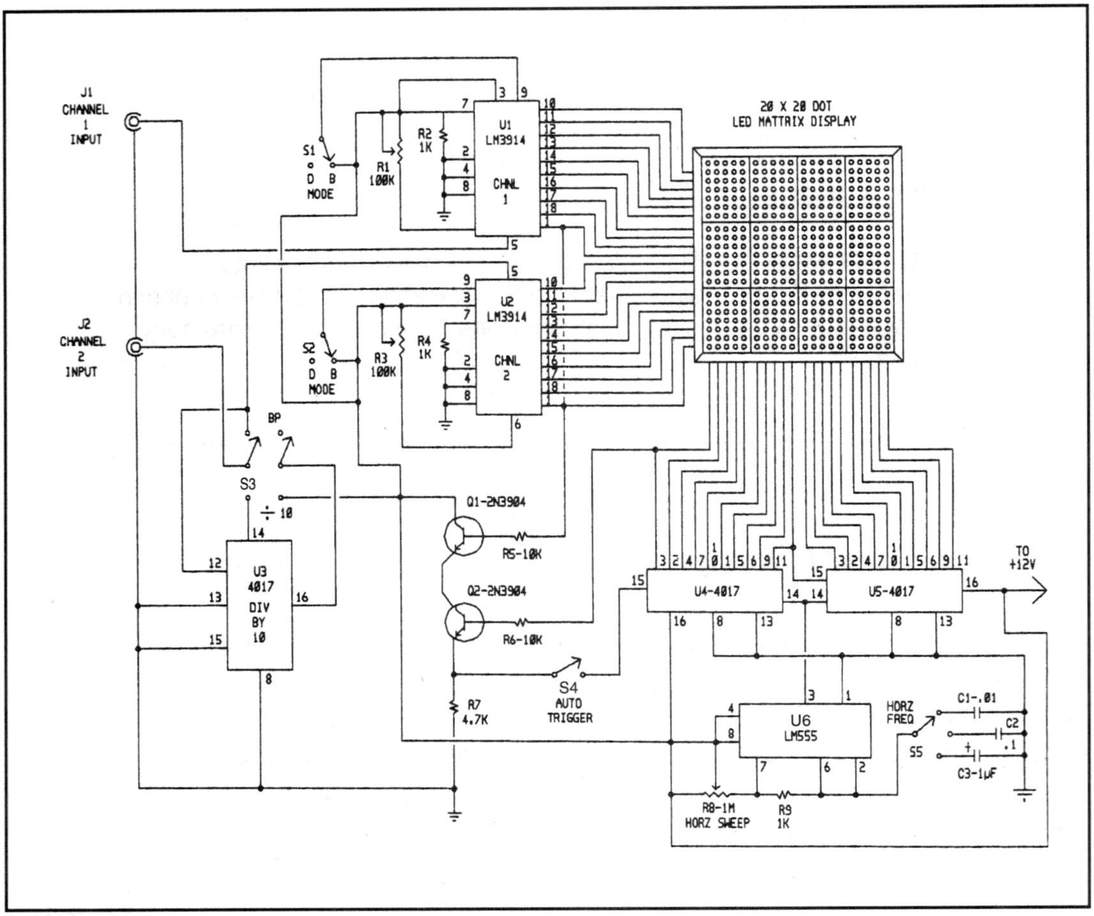

Figure 12-1. Solid-state oscilloscope.

Finally, a power source is required to keep all the goodies inside hopping. As stated, conventional CRT scopes use a lot of high voltage, but our solid-state version requires only a regulated 12 volts.

Some other bells and whistles have been added to this design, but we will get to that later. For the moment, let me describe how this system functions.

Referring to *Figure 12-1*, achieving vertical gain is the job of an old standby, the LM3914 bar graph display driver. With this chip, ten separate outputs transition from a high state to a low state as the input voltage increases. The 3914, or another from that series, is frequently used as a volume unit (VU) meter with audio equipment. Two such LM3914s (U1 and U2) are employed, and each is wired as a separate vertical gain circuit. This results in a *dual-trace* arrangement in which two individual signals can be introduced and evaluated at any given time.

The display consists of twelve 5 x 7 dot-matrix LED units which provide a *20 row x 20 column* resolution (total of 400 diodes). Actually, there are 21 rows, but either the top or bottom row is left unused.

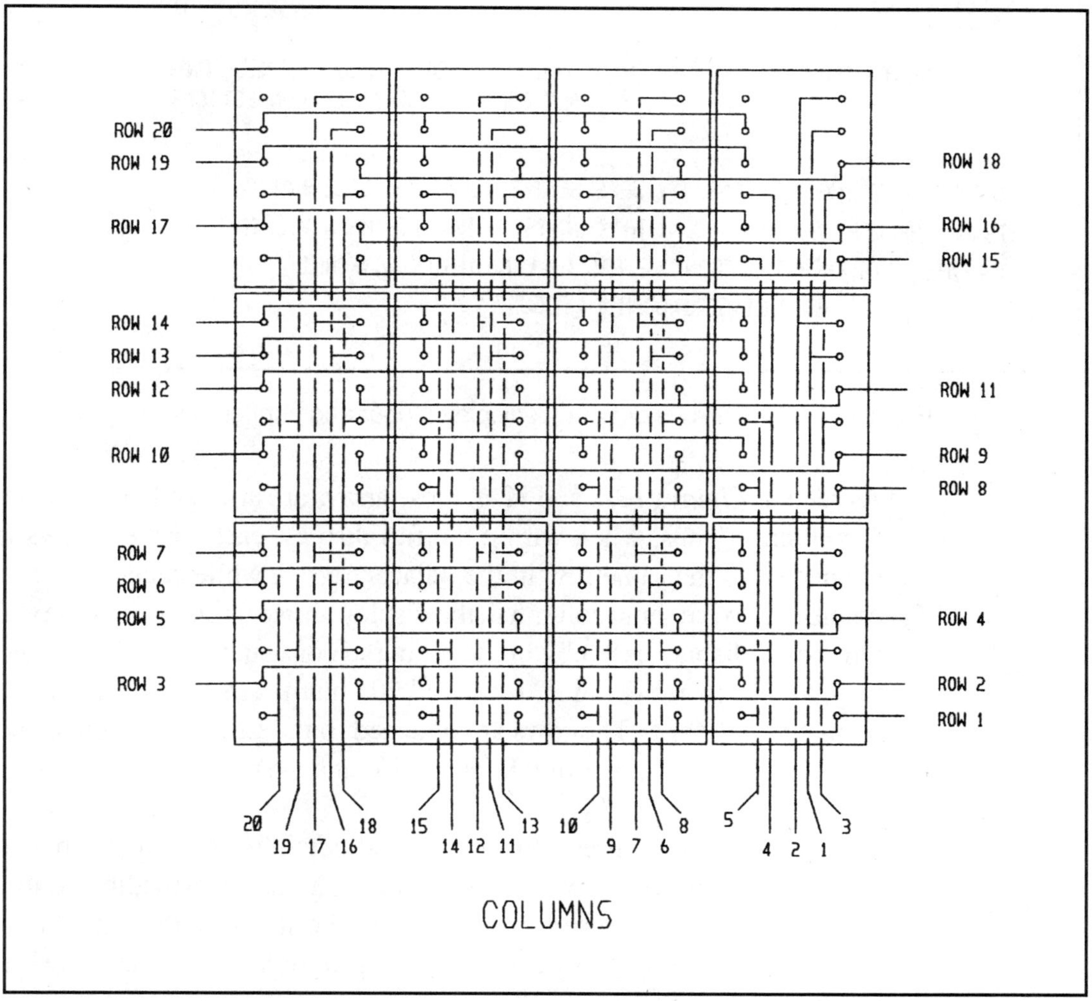

Figure 12-2. LPT747R dot matrix display wiring diagram (solder side).

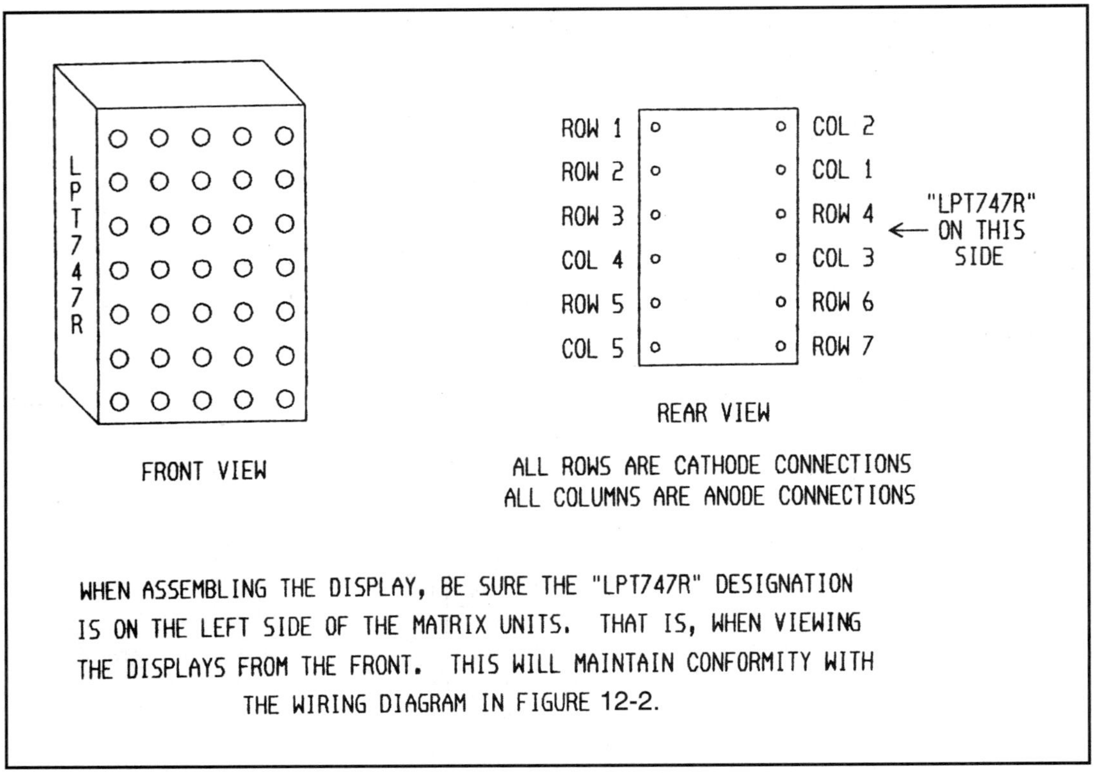

Figure 12-3. LPT747R dot matrix display pinout and configuration.

As seen in *Figures 12-2* and *12-3*, these arrays are arranged in 7 rows by 5 columns. Thus, when wiring the full display, each of the 7 rows is connected to the same row in the adjacent array, while each column is connected to the same column in the display above or below it. Relax! I will cover display assembly in detail in the construction section. Anyway, all of this results in 20 rows and 20 columns that are wired to the main circuit board. The rows are attached to the LM3914 outputs, and the columns go to a pair of CD4017s (U4 and U5).

Since each LED requires a low on the cathode in order to light up, and a high on the anode, the two LM3914s do their part by providing 20 low outputs; that is, low when activated. As the input voltage increases, more of the 3914 outputs activate, which progressively elevates the image on the display.

Now we need some form of *sweep* circuitry to draw the image out and across the display. Without this, you would simply get the same, familiar single-line annunciation available in audio VU meters.

Another requirement of the sweep circuit is to provide high or positive levels. This is needed to satisfy the LED anodes. So, an old friend immediately comes to mind: the CD4017 decade counter/divider (a.k.a., the Johnson counter). With this beauty, 10 outputs consecutively go high in step with the signal from a clock source. By cascading the two chips, we now have 20 outputs going high one at a time. Since this is sequential, the 4017s move the image across the display.

While they are doing that, adjust the 3914's height according to the input voltage; voila, you have a waveform. Vertical gain adjustment affords control of this function, and a combination of horizontal frequency and sweep rate determines the speed at which the trace is displayed (how fast it moves across the display).

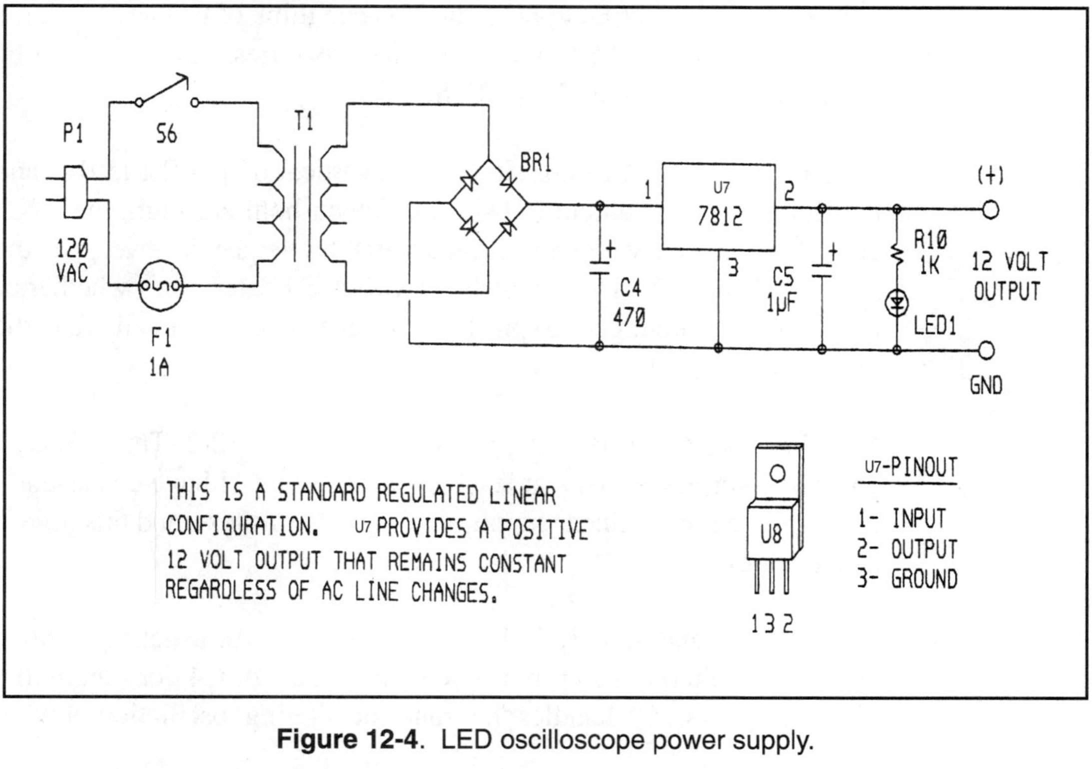

Figure 12-4. LED oscilloscope power supply.

The clock source can be any of a number of different circuits. For example, you could use digital oscillators constructed from inverter or NAND gates. However, in the prototype, I selected another old friend: the 555 oscillator/timer IC (U6).

This system is easy to adjust, and relies on commonly available support components. The 555 is highly reliable, but I do suggest using the *NE* version, not the *C* or CMOS. For whatever reason, I have always had trouble with the low-power ICs when applied to this circuit.

Glancing at *Figure 12-1*, you will notice two transistors and three resistors (Q1, Q2, R5, R6, R7). These are arranged as an *AND* gate. Why, you ask? Well, what we have here is the *synchronization* circuit; that is, a method to get the display started in the right place each time. If this were not present, there is no telling what would come up on the display. Trust me, you need this circuit.

If you are familiar with AND gates, then you know that both inputs have to be high for the output to be high. You can think of this as two single pole-single throw (SPST) switches wired in series. Both have to be closed for the current to flow. Right?

To get the synchronization, tap the first vertical output (LM3914) and the first horizontal output (CD4017). When both are high, the AND gate output is high, which resets (and starts) the sequential sweep via the first 4017's pin 15. The *auto trigger* switch S4 determines whether or not the AND output goes to pin 15. This can be used to manipulate the display, as you will see later.

Next, let's look at the power supply, as seen in *Figure 12-2*. This shouldn't come as a surprise to you, at least not at this point. I have consistently employed regulated linear supplies throughout this text, and this time is no exception.

Yes sir, one handy-dandy 7812T configuration. As usual, the bridge rectifies the alternating current (AC), and capacitor C4 does the initial filtering honors. C5 handles the transient filtering, oscillation prevention and bypass, while resistor R10 and LED1 provide the *on* indication. Just as simple as always, and more than enough power for our design, without having to deal with *high voltage*. In all fairness, I realize that the CRTs have to have high voltage, but it seems like every time I delve inside such a device, it bites me.

I mentioned some bells and whistles, so let's get to that part. First, the LM3914 can be set to deliver two styles of display. There's the *bar* mode, which we have already examined (all LEDs remain *on* as the voltage increases and the outputs activate). However, the chip can also produce a *dot* display, in which only the most recently activated output is low. In this scenario, the previously activated outputs return to high, extinguishing the LEDs connected to them. The advantage of both modes will become apparent when we get to the operational section, though you may have already guessed what this is all about.

Switches S1 and S2 accomplish the transition by taking the LM3914's pin 9 either high (bar mode) or floating (dot mode). Taking pin 9 low will also result in a dot mode, but why bother with that when floating it works just as well?

Since we are dealing with LM3914s, the input frequency response is limited to about 100 kilohertz. This level is often more than satisfactory, but if you want to measure higher frequencies, you need a way to increase the response.

Needless to say, we can't go in and change the architecture of the chip, so I added a third CD4017 configured as a *divide-by-10 counter*. This allows the input signal to first be divided by 10 before introducing it to the system. You'll note that this feature is only present on the second channel, and is controlled by switch S3. If desired, you can incorporate it in both channels by using a second 4017 and switch.

S3 is a DPDT unit, with one side (A) determining the destination of the input signal, and the other half (B) applying power to U3. The *A* section routes the signal either directly to the 3914 input or through the divide-by-10 counter, then to the 3914 input. The *B* section pow-

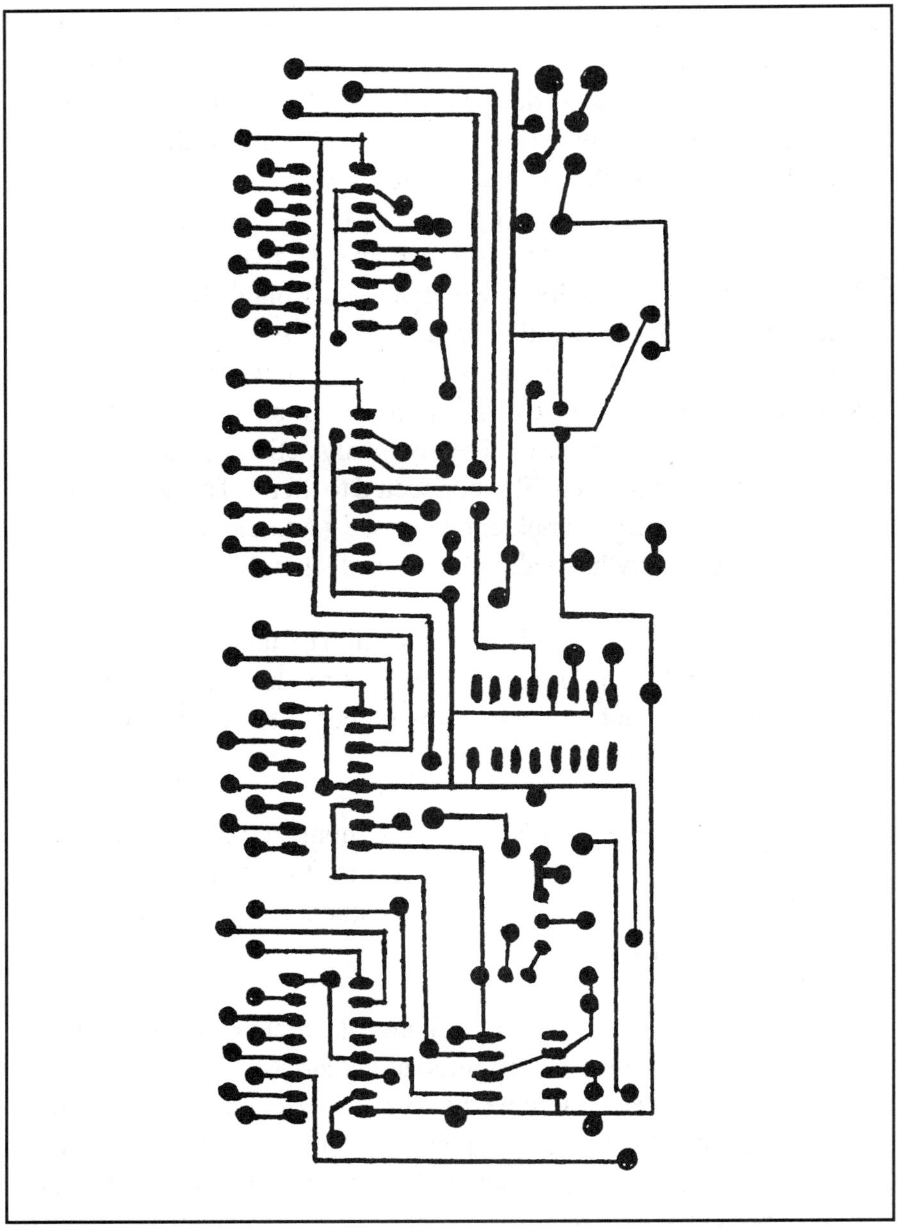

Figure 12-5. Solid-state oscilloscope foil.

ers U3 only when it is in use. This not only saves on energy consumption, it also prevents the 4017 from causing input interference.

For those of you familiar with the interior of CRT scopes, I'm sure you will agree that LED oscilloscope theory is substantially less complex. For those of you who are not, take our word for it.

Construction

I highly recommend a printed circuit board (PCB) for this one. As it is, you are going to have enough point-to-point wiring with the panel components and display. No sense in making more work for yourself.

To that end, I have graciously included *Figures 12-5* and *12-6* which respectively provide the PCB foil pattern and parts placement. The board is single-sided and simple to lay out and etch. With it, the project really will be easier to construct.

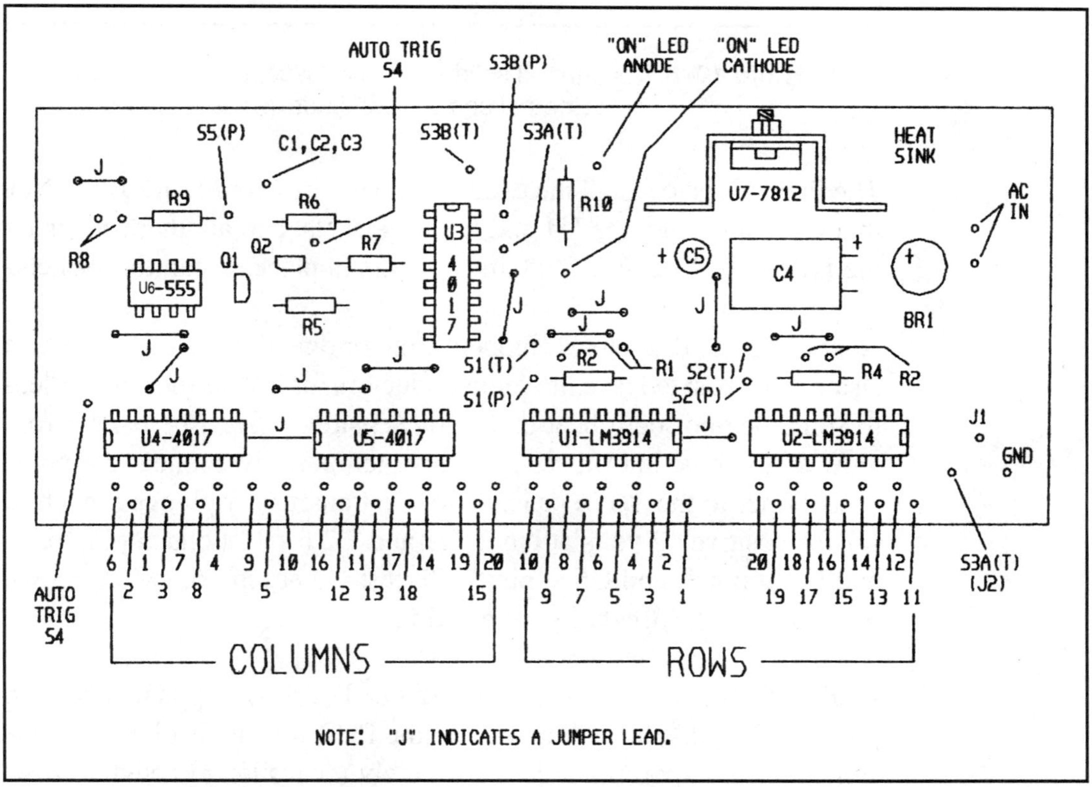

Figure 12-6. LED scope parts placement.

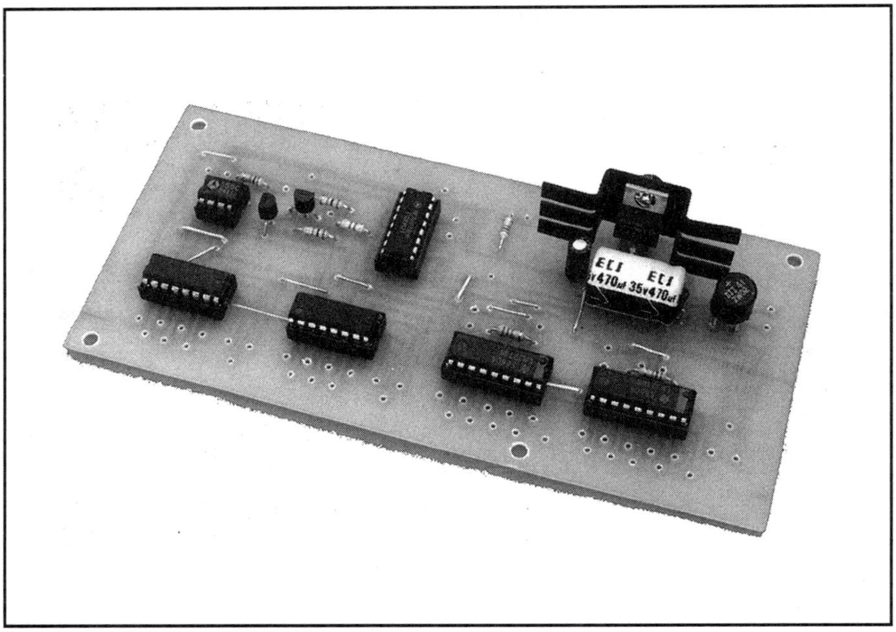

Photo 12-1. The stuffed solid-state oscilloscope board without external component leads.

The first task is to install the twelve jumper wires per *Figure 12-6*. Note that one runs under the U4 and U5 IC sockets, and another runs under the U1 and U2 sockets. With the jumpers in place, install the sockets.

Next come the discrete components, the bridge, regulator U7 and transistors. It's best to leave the semiconductors for last, as they are (at least to some degree) both heat and static sensitive. See *Photo 12-1* for a view of the board at this stage. Now, temporarily connect the power transformer to the *AC in* pads. With the transformer plugged in, check U7 for positive 12 volts at pin 2, U1 and U2 for 12 volts at pin 16, U4 and U5 at pin 3, and U6 at pin 8. To test U3 at pin 16, switch S3 will have to be in the divide-by-10 position.

If all is well, disconnect the transformer. The next step is to solder appropriate lengths of hookup wire to the PCB for the display and panel components. Appropriate lengths merely means long enough to reach whatever it is you are trying to reach.

Once completed, set the main board aside and turn your attention to the display. This part is going to be a bit tedious, as over 350 solder joints have to be made. The best advice is to go slowly; patience is the key.

For the prototype, I assembled the display arrays on solder ringed-hole perf board, and DID NOT use IC sockets. These units need to be perfectly flush with each other, so the end product is a 2" (50 mm.) by 2-1/8" (53 mm.) solid square of LED arrays. Unfortunately, sockets will throw off the spacing.

Using tinned bus wire, start with the rows, then do the columns. With rows, each *row 1* pin is connected to the next array's *row 1* pin; *row 2*

Photo 12-2. The completed display board for the oscilloscope. Note the connection leads on all four sides.

to the next *row 2*, and so forth. The same applies to the columns; all *column 1* pins are connected to a single line, all *column 2* pins to another, etc. In the end, you will have 20 of each, with the rows running on each side of the display board, and the columns on the top and bottom (see *Photo 12-2*).

It is advisable to check each individual row and column after soldering. Be sure that all four *row* LEDs light up, and all three *column* LEDs light up. If not, you probably have a cold joint or a just plain ol' nonjoint. It is far easier to correct these mistakes as you go along than to try to remedy the errors after the display is completed.

Once the display is finished, run a line-by-line check on all 40 leads. I did this with a 3-volt battery supply. Connect the negative terminal to

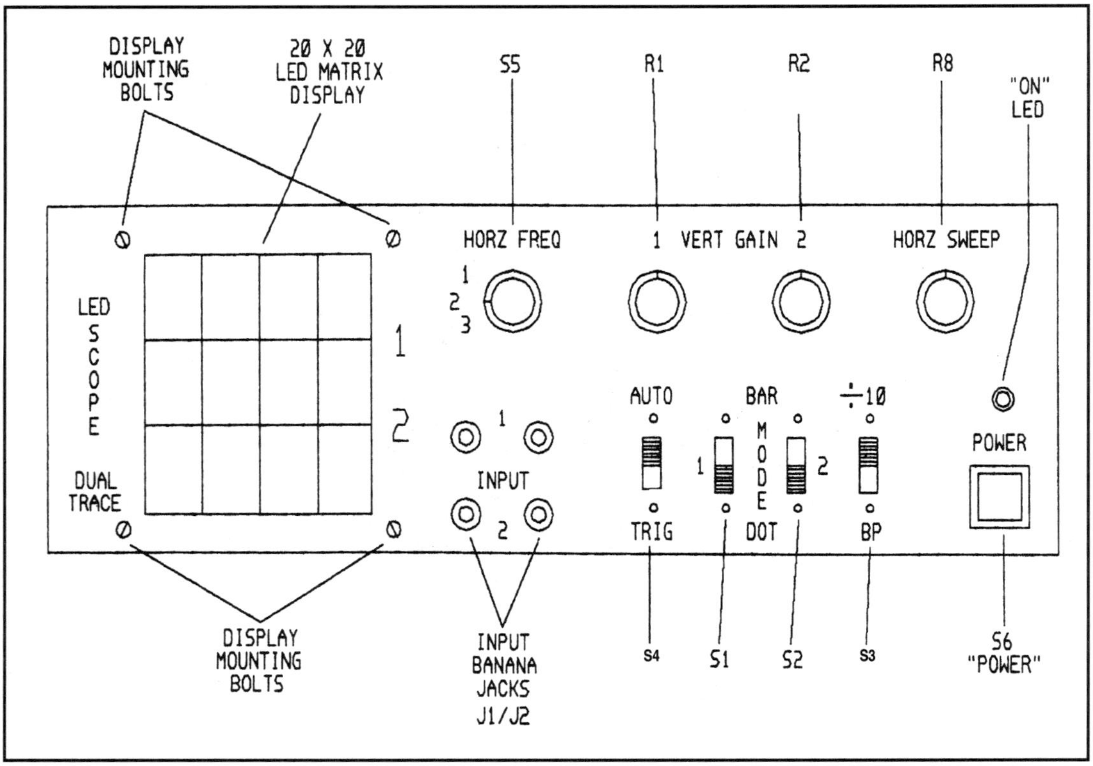

Figure 12-7. Solid-state scope template.

the rows and touch each column lead with the positive line. What you are looking for is a single LED-illuminating *dot*. If you get more than one, you have a short. Go back and trace the problematic lead to each connection, and the trouble will eventually show up. Also, solder bridges are easily accomplished with spacing this tight, so look for these as well.

When everything is right, give all of the interconnecting bus wire leads a coat of clear nail polish. This helps prevent accidental shorts while handling the display board.

Congratulations! You have just completed the hardest part of this project. Take a break. You deserve it!

Okay, that's long enough. Let's move on. You may set the display to one side for the moment.

Now is the time to prepare the case. Any suitable enclosure will work, but I recommend a 10" (256 mm.) by 7-1/2" (192 mm.) by 3-1/8" (79 mm.) instrument case. Internally this is overkill (see *Photo 12-5*), but it does provide badly-needed panel space.

Cut and drill the necessary openings in the front panel, then spray paint it with a light color. The back panel needs only holes for the line cord and fuse holder.

I painted the shell sections of the enclosure medium blue; however this can be left natural if the color appeals to you. What you do here is strictly a matter of personal preference.

When the front panel coat is dry, label it with dry-transfer lettering. *Figure 12-7* shows a template for the prototype, which can be used as a guide. Again, personal preference prevails. Once lettered, spray a coat of clear enamel on the panel to protect the labels. As always, keep this coat light. While the paint is curing, it's time to finish the wiring.

Back at the board, connect all the panel-mounted components to their leads. The banana jacks used for inputs will have to go on the front panel before they can be soldered to the hookup wires.

Now comes the display. You have 40 leads to connect to the board, and they all have to go to the correct display board point. Be careful! This is not the time to get sloppy. I did, and I regretted it. Also, be as gentle as possible with the display while you work with it. With this many leads, it is very easy to break a few, resulting in the time-consuming and irritating task of going back and repairing the damage (*Photo 12-3*).

With all the display leads connected, fasten the display to the front panel with four bolts and standoffs. Next, if you haven't already done so,

Photo 12-3. The back side of the display board showing the labyrinth of connections.

mount the panel potentiometers, switches, jacks and *on* LED. Last, slip the front panel into place and secure the main board to the bottom half of the case.

If you've come this far and haven't given up in total frustration, the unit is done. Install the ICs, and you're ready to test the system.

Testing and Operation

Testing is fairly simple. Plug in the unit, set the horizontal frequency to *2* and the horizontal sweep to mid-scale, and hit the power switch. IT'S ALIVE, IT'S ALIVE! Uh... sorry. Anyway, you should see two lines of vertical dots moving across the display. If not, double check the power supply and wiring. If the two vertical lines are present but not moving, this indicates a problem with the clock circuit.

Try switching the horizontal frequency to either *1* or *3* and adjusting the horizontal sweep. Also, as I stated earlier, the C555 or CMOS timer chips have given me problems in the past, so you may want to examine the one in the oscilloscope. Try replacing it with a NE555. This often cures the ill.

Next, increase the sweep control. The display will advance from faster-moving blocks of illuminated LEDs to a solidly-lit display. Actually, only one line of LEDs is on, but due to persistence of vision, the entire display will appear to be on.

Now, connect a signal source to the input(s). This can be the function generator we built in Chapter 10, or any other square, sine or triangular signal. By adjusting the vertical gain, horizontal frequency and sweep, a waveform representative of the input will appear. You will note that as the source frequency changes, the space between the waveform cycles will also change (closer for higher frequency, greater spread for lower). Also, careful adjustments of the horizontal sweep or deactivating the auto trigger will freeze the form on the display.

Mode switches S1 and S2 will select either the *dot* or *bar* pattern. With the dot mode, a more traditional *single-line* waveform will be seen, while the bar mode will fill in the form. The latter can make the sign pattern easier to distinguish, particularly with square waves.

Since this scope is of the dual-trace or 2-channel variety, two separate signals can be analyzed at the same time (*Photo 12-4*). This, of course, is very handy when signal comparison is needed.

Incidentally, by cascading the two LM3914 chips, you can convert this system to a single-channel (trace) scope. This will improve the resolution, but (naturally) only one input can be viewed. Something to think about as a possible modification.

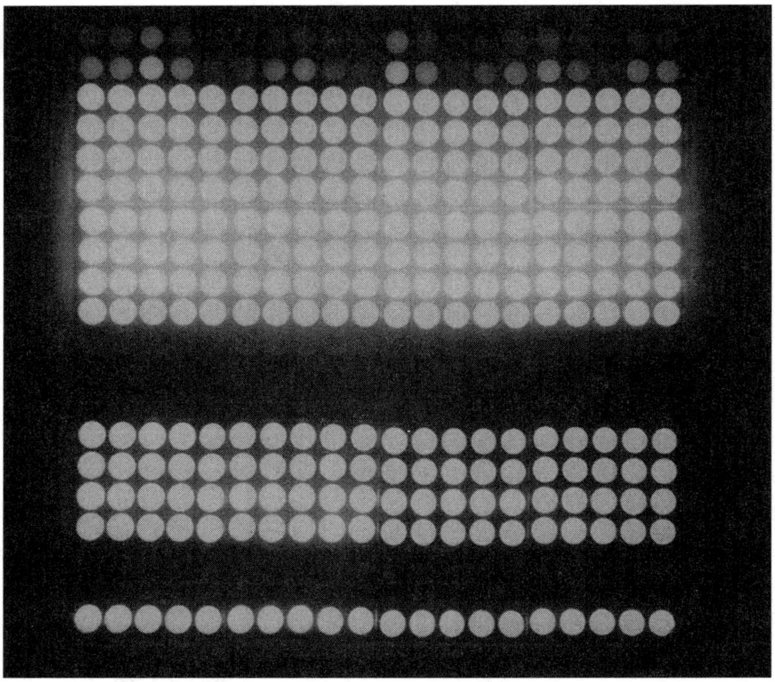

Photo 12-4. A shot of the dual-trace capability of the scope. The top is the bar mode while the bottom is the dot mode.

Once you have tackled the general operation, try introducing specific signals; that is, specific in terms of frequency and voltage. By leaving the controls in one position, you will be able to establish vertical height and horizontal spacing for various input signals at these settings.

Different control assignments will yield different images, and this will allow you to take measurements with the scope. If desired, a grid can be laid out on clear plastic as a guide for the measurements.

On a last note, as with conventional CRT scopes, any waveform can be displayed by this device. Well, just about any waveform. This, of course, adds to the versatility of the system, so don't be afraid to try a variety of signals. I think you will be pleased with the results.

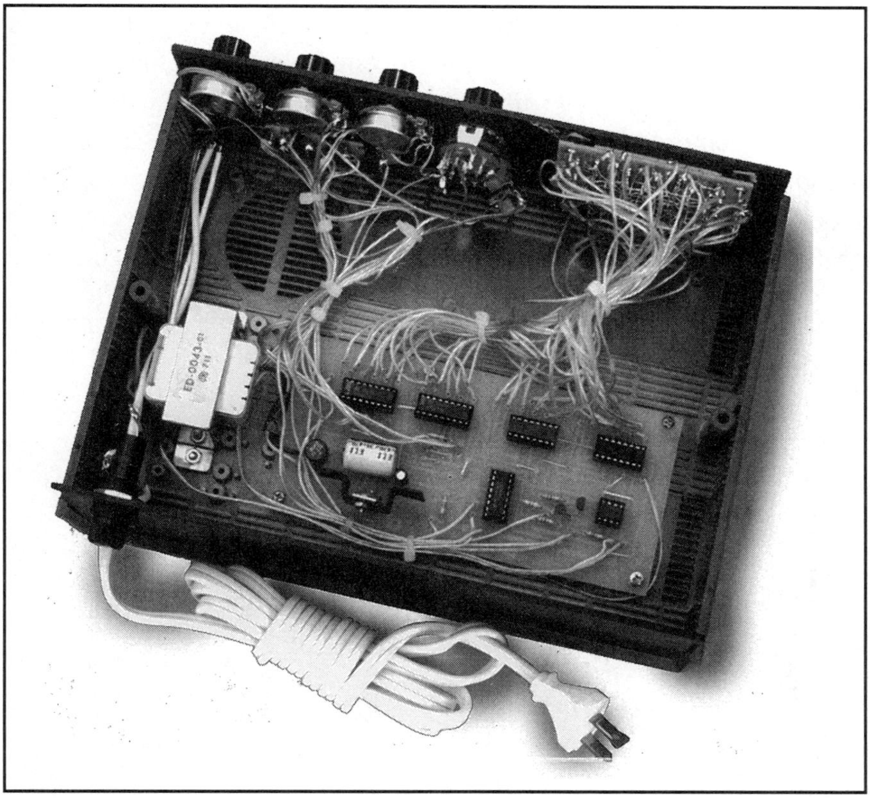

Photo 12-5. Internal shot of the completed LED oscilloscope. Note the wires leading to the front panel.

Conclusion

Well, what do you think? Nifty, ain't it? Really, this solid-state oscilloscope is a fun project to build, and you end up a very practical piece of test equipment. Once you get past wiring the display (*Photo 12-3*), the rest is a snap.

In addition to lower power requirements, this technology provides an excellent alternative to a present-day industry standard; the CRT oscilloscope. While CRT devices will probably always have a place in electronics, the solid-state approach will, in the future, certainly narrow the breadth of that place.

In reality, solid-state scopes have been around for quite awhile. Over the years, a variety of different concepts have been developed, and many have functioned well. However, it has only been lately that manufacturers have given solid-state displays a chance to prove their worth. Down the road, I venture to say a lot more will be heard from this technology. Already, computer folks have discovered solid-state devices for laptop models. Miniature televisions with LCD screens also have become very popular. All in all, the cathode ray tube is rapidly becoming a dinosaur;

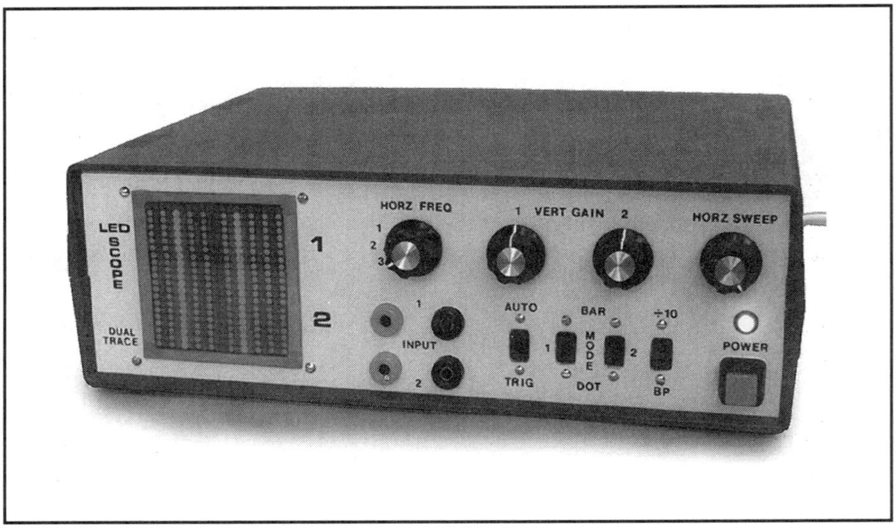

Photo 12-6. The completed solid-state oscilloscope in operation.

probably never extinct, but a dinosaur nonetheless. In all fairness, the CRT has served us well. We wouldn't have modern television without it, and various versions of the conventional oscilloscope benefit many aspects of our lives. But this doesn't mean we shouldn't look toward new and better ways to do the job.

This, of course, brings us to our LED scope (*Photo 12-6*). While improvements are always on the horizon, this device provides an inexpensive and practical way to do many tasks once possible only with the CRT scopes. Also, this project lends itself eagerly to modification, and in that lies progress, which isn't always bad.

So, have fun with the LED oscilloscope. Once completed and operational, you will soon learn to rely on it as a major component of your test lab. For a little entertainment, try connecting it to an audio amplifier/microphone assembly. Watching the display respond to your voice, music and other surrounding sounds should furnish ample amusement.

Conclusion

Well, we've come to the end of the trail. Within these pages, I truly hope you have found what you need or want, or both. While your test bench doesn't have to resemble Bell Laboratories, it is nice to have these devices on hand. They will all make your work much easier, and that should equate to FUN!

Among these test equipment marvels, I have my favorites; the solid-state oscilloscope, the frequency counter and the capacitor tester, just to name a few. But I'm always grateful to have each of them available.

When digital electronics is the subject, the logic probe is an immense help. The capacitor, transistor and integrated circuit testers help keep the parts drawers full and the components honest.

The digital counter, function generator and multimeter will prove their merit surprisingly often, while the bench power supply and signal injector/tracer are staples for the experimenter. Also, you will wonder how you ever existed without the prototype lab.

All in all, every device in this book will assist you in your quest of expanding your knowledge of electronics, no matter how simple they may seem at the outset. In simplicity lies one of the truly remarkable aspects of this field: terrific things can come in small packages as well as big ones.

Also, just because it's electronics doesn't mean it has to be complicated. With a solid understanding of electronics basics, most individuals find that this arena of science is quite manageable. Even (should I say it again?) FUN! The test instruments in this book can be of tremendous service in helping you achieve this basic understanding. That and good old-fashioned hands-on experience. At the risk of repeating myself, it has been my observation that few approaches to learning this stuff are surpassed by rolling up your sleeves and getting your hands dirty. This is perhaps a long-winded way of saying, "Learn by doing." But one of the truly treasured aspects of electronics is that it always has something new to offer. New chips hit the market almost daily, and technology improves so rapidly it's impossible to keep up with it.

All of this is exciting and challenging, but at times it can also be confusing. However, a well-equipped test bench will help you sift through the new information and develop the new hardware. This, of course, brings us back to the purpose of this text: to provide you with a well-equipped test bench.

So, by now, you should have all these projects up and running, right? What, you don't? Well, keep working on it, because when you finish, you're gonna kick... well, you know, and take names when it comes to dealing with everything that electronics has to offer.

Seriously, in any type of work, there is no substitute for the right tools. Electronics is no different. Just as a soldering iron and wire cutters can help you produce successful projects, this equipment will vastly support your creativity while developing prototypes. Additionally, the ability to analyze circuit operation ensures faster, more precise troubleshooting.

Okay, I've said my piece. Now it's your turn. Build these gems and let me know how much you like them. Shine on, gang!

Conclusion

Appendix
Source List

The following list represents the companies I have used in the past for parts, supplies and other items needed to build the projects in this book. Of course, you may know of others.

For each source, I have included the address, as many phone numbers and other means of contacting them as I could dig up, and a brief list of what they carry; at least, what they carry related to these projects. Hopefully, this list will be helpful.

ALL ELECTRONICS CORPORATION
P.O. Box 567
Van Nuys, CA 91408-0567
Phone: (800) 826-5432
Good line of surplus gear as well as useful parts. Source for 6PDT pushbutton arrays for Chapter 9.

ALTRONICS
2300 Zanker Road
San Jose, CA 95131
Phone: (408) 943-9773
Fax: (408) 943-9776
Many older ICs, enclosures, switches and other parts. Possible source for pushbutton switch arrays.

AMERICAN SCIENCE & SURPLUS
3605 Howard Street
Skokie, IL 60076
Phone: (847) 982-0870
Fax: (800) 934-0722
Tremendous source of surplus electronic parts, hardware, etc.

B.G. MICRO
P.O. Box 280298
Dallas, TX 75228
Orders: (800) 276-2206
Tech: (972) 271-9834
Fax: (972) 271-2462
Internet: http://www.bgmicro.com/
E-mail: bgmicro@bgmicro.com
Good source for surplus equipment and parts.

THE C-THRU RULER COMPANY
Post Office Box 356
Bloomfield, CT 06002
Orders/Info: (860) 243-0303
This is the source for dry-transfer lettering if you cannot find it locally. Excellent lettering.

CIRCUIT SPECIALISTS, INC.
P.O. Box 3047
Scottsdale, AZ 85271-3047
Orders: (800) 528-1417
Info: (602) 464-2485
Fax: (602) 464-5824
Internet: http://www.cir.com
Good line of ICs, components, hardware and other supplies.

DC ELECTRONICS
Post Office Box 3203
Scottsdale, AZ 85271-3203
Phone: (602) 945-7736
Orders: (800) 467-7736 or (800) 423-0070
Fax: (602) 994-1707
Good line of specialized ICs and transistors.

DEBCO ELECTRONICS
4025 Edwards Road
Cincinnati, OH 45209
Orders: (800) 423-4499
Info: (513) 531-4499
Fax: (513) 531-4455
Carries the capacitance adapter kit used in Chapter 7.

DIGI-KEY CORPORATION
701 Brooks Avenue South
Thief River Falls, MN 56701-0677

Orders: (800) 344-4539
Info: (218) 681-6674
Fax: (218) 681-3380
Internet: http://www.digikey.com
Large catalog of recent ICs, parts, coils, switches, discrete components, etc. Carries caps for pushbutton switches.

ELECTRONIC GOLDMINE
Post Office Box 5408
Scottsdale, AZ 85261
Phone: (602) 451-7454
Fax: (602) 661-8259
Internet: http://www,goldmine-elec.com
Surplus house that carries parts, hardware, cases, etc.

JAMECO ELECTRONIC COMPONENTS/COMPUTER PRODUCTS
1355 Shoreway Road
Belmont, CA 94002-4100
Phone: (415) 592-8097
Order: (800) 831-4242
Fax: (800) 237-6948
Internet: http://www.jameco.com
Full-service parts distributor for parts, cases, hardware, etc. Also carries the digital meter for the Chapter 7 project.

JDR MICRODEVICES
1850 South 10th Street
San Jose, CA 95112-4108
Phone: (408) 494-1400
Order: (800) 538-5000
Fax: (408) 494-1420
Internet: http://www.jdr.com
Full line of parts, cases, hardware and other useful items. Source for capacitance adapter and meter in Chapter 7.

MARLIN P. JONES & ASSOCIATES, INC.
P.O. Box 12685
Lake Park, FL 33403-0685
Phone: (800) 652-6733
Fax: (800) 432-9937
Internet: www.mpja.com
E-mail: mpja@mpja.com
Extensive inventory of parts, hardware, transformers, etc. Also has the capacitance adapter kit and meter from Chapter 7.

MCM ELECTRONICS
650 Congress Park Drive
Centerville, OH 45459-4072
Phone: (800) 543-4330
Good line of useful parts, hardware, etc., for projects.

MENDELSON ELECTRONICS COMPANY, INC. (MECI)
340 East First Street
Dayton, OH 45402
Orders: (800) 344-4465
Fax: (800) 344-6324
Carries parts, hardware and other surplus items.

MOUSER ELECTRONICS
National Circulation Center
2401 Highway 287 North
Mansfield, TX 76063-4827
Phone: (800) 346-6873
Internet: http://www.mouser.com
E-mail: sales@mouser.com
Extensive line of parts, chemicals and PC board materials.

RADIO SHACK
Local Stores
Tandy Corporation
Ft. Worth, TX 76102
Handles a limited line of parts and hardware. Good source for switches and cases.

RAINBOW ELECTRONICS, INC.
6227 Coffman Road
Indianapolis, IN 46268
Info: (317) 291-7262
Fax: (317) 291-7269
Manufacturer of the capacitance adapter used in Chapter 7.

RAMSEY ELECTRONICS
793 Canning Parkway
Victor, NY 14564
Phone: (716) 924-4560
Orders: (800) 446-2295
Fax: (716) 924-4555
Internet: http://www.ramseyelectronics.com
Carries the capacitance adapter for Chapter 7.

TECH AMERICA
P.O. Box 1981
Fort Worth, TX 76101-1981
Orders: (800) 877-0072
Fax: (800) 813-0087
Tech Hot Line: (800) 876-5292
Cust. Serv.: (800) 613-7080
A tremendous catalog full of parts, hardware, enclosures and just about everything you need.

TIMELINE INC.
2539 West 237th Street
Building F
Torrance, CA 90505
Orders: (800) 872-8878
Tech. info: (310) 784-5488
Fax: (310) 784-7590
Good line of surplus items, including chips, LCD displays and other useful stuff.

Glossary

ALTERNATING CURRENT (AC): Nonpolarized power that is constantly changing back and forth between positive and negative. An example is your household current.

AMPERAGE: The unit used to measure the amount of current present or the amount needed for a circuit.

AMPLIFIER: Any device that increases the strength of a signal. It can be as simple as a single transistor or as complex as dozens of semiconductors.

AMPLITUDE MODULATION (AM): A form of modulation in which the base frequency stays the same, but the strength or amplitude (height) of the cycles changes.

ANALOG: Any electronic circuit that utilizes variable information. In other words, the input values can be something besides 0 or 1.

ANODE: The positive terminal of a diode or other polarized component.

ANTENNA: Any device that either radiates a signal or pulls in signals.

AUDIO: Applies to a signal that is within the hearing range of the human ear. With most people, this tops out at about 20 kilohertz.

BARDEEN, JOHN: One of three scientists credited with inventing the transistor in 1947.

BAR GENERATOR: A device designed to create bar patterns on a television set. The bars can be color or B/W. Many units also produce other patterns.

BIAS: Refers to the voltage applied to any particular part of a circuit.

BRATTAN, W.H.: One of three scientists credited with inventing the transistor in 1947.

BROADCAST: One term for a radio or television program, or the process of sending data by radio waves.

CABLE: Wire, or wire assemblies used to carry signals or power within a circuit, or between circuits.

CAPACITOR: An electronic component that has the ability to store a charge and block DC voltage/current.

CATHODE: The negative terminal of a diode or other electronic component.

CATHODE RAY TUBE (CRT): A specialized vacuum tube that uses an electron gun and a phosphorescent coating to produce an image on its screen. Often referred to as the *picture tube*.

CLOCK: Another term for a timing circuit.

CHIP: A semi-slang term used to designate an integrated circuit.

COIL: A device that is made up of one or more turns of wire which creates a magnetic field when electricity is passed through it.

COLOR BURST: The frequency established by the NTSC as the standard from which all TV sync frequencies are derived. The frequency, 3.579545 megahertz, is often used as a reference signal.

DIGITAL: The opposite of analog, these signals are designated as either *0* or *1*. There is no in-between.

DIGITAL COUNTER: A device that uses digital electronics to count signal pulses.

DIODE: An electronic component that changes alternating current to direct current. This process is called *rectification*.

DIRECT CURRENT: Voltage that has polarity; in other words, it has a negative side and a positive side, which do not change. Batteries are a good example.

ELECTRON GUN: A combination of a filament, grids and plates that produces a stream of electrons when properly biased. This is the heart of a cathode ray tube.

FARAD: Named after German mathematician Michael Faraday, this is the unit of measurement for capacitors.

FEDERAL COMMUNICATIONS COMMISSION (FCC): The U.S. government agency charged with supervising the airwaves. It does its job by issuing licenses to qualified broadcasting companies and individuals.

FREQUENCY: The wavelength at which a transmitter operates. It also applies to oscillators and other repeating functions.

FREQUENCY COUNTER: A specialized digital counter that only counts during a specified time period or window. This results in an accurate count of the frequency of the input signal.

FREQUENCY MODULATION (FM): A modulation system in which the amplitude of the cycles remains constant, while the repetition of the cycles changes.

FUNCTION GENERATOR: An electronic circuit that is capable of producing waveforms, usually sine, square or triangular waves. Some units provide adjustable frequency outputs.

GIGA: Pronounced "jig-a," the term means one thousand millions, or one billion. A gigawatt is one billion watts.

GIGAHERTZ (GHz): Pronounced "jig-a-hertz," one gigahertz is equal to 1,000 megahertz. This is also known as the microwave region.

GROUND PLANE: A type of antenna that utilizes a center element for transmitting/receiving and grounded elements (usually at 90 degree angles from the center rod) as reflectors.

HENRY: Named for American scientist Joseph Henry, this is the measurement for inductors, or coils.

HERTZ: The designation for frequency named after German physicist Heinrich Hertz (not the rent-a-car people).

HORIZONTAL: In graphic displays, back and forth movement, as opposed to up and down.

HORIZONTAL SWEEP: The process of making one complete scan across a display. Can also refer to the time required for the scan.

IMAGE: The pattern seen on the display device of an oscilloscope.

INDUCTOR: An electronic component consisting of turns of wire that set up a magnetic field. Important to tuning and frequency control.

INFRARED: Invisible light waves below the visible red spectrum; usually ending around 750 nanometers.

INTEGRATED CIRCUIT (IC): Any of a huge number of semiconductor packages that contain entire circuits. ICs have become an accepted standard within the entire electronics industry.

INTEGRATED CIRCUIT TESTER: A device that is used to test and identify integrated circuits and their pins.

KILO: A prefix that indicates one thousand times the unit in question. One kilowatt is a thousand watts.

KILOHERTZ (kHz): One kilohertz is equal to 1,000 hertz. Not seen in the television bands.

LASER: A light-emitting device that produces a narrow beam through collimation and coherency. The name is an acronym for *Light Amplification by Stimulated Emission of Radiation.*

LIGHT-EMITTING DIODE (LED): A semiconductor device that glows when power is applied to its electrodes. These are specialized diodes usually made with a gallium arsenide substrate.

LIQUID CRYSTAL DISPLAY (LCD): A type of display that uses special liquid crystals to form an image. These crystals are normally clear, but when power is applied, they twist and exhibit color.

LOGIC PROBE: An electronic test device that detects the status of a signal. It can tell you if the signal is high (1) or low (0). Sometimes it can tell you if the signal is pulsing.

MEGA: A prefix used to denote one million times a given quantity. One megawatt is one million watts.

MEGAHERTZ (MHz): One megahertz is equal to 1,000 kilohertz, or one million hertz. This is the unit in which all TV bands are measured.

MICROPROCESSOR: A specialized integrated circuit that can contain millions of transistors. Microprocessors are the brains of computer systems, and are very useful for control purposes.

MILLI: A prefix that denotes one thousandth of a quantity. One millivolt is 1/1000 of a volt.

MONITOR: Any device used to display a video signal. It can use a cathode ray tube or an LCD array.

MULTIMETER: A metered instrument, either analog or digital, that can measure several different quantities. The most common systems read ohms, volts and amps, but some also provide frequency and capacitance measurements.

NATIONAL TELEVISION STANDARDS COMMITTEE: Abbreviated the NTSC. The government group that worked with the television industry to establish a universal broadcasting standard. While not used worldwide, it is the standard for North America, Japan and a few other areas. The NTSC format is based on 525 lines and a 60-hertz vertical speed.

OHM: Named after German mathematician George Simon Ohm, this is the unit of measurement for resistors and some other electronic components.

OSCILLATOR: An electronic device that produces an alternating output. Some common oscillators are transmitters and clock circuits.

OSCILLOSCOPE: A piece of test equipment that contains a specialized cathode ray tube or solid-state display. Waveforms can be displayed to measure frequency and power.

PHASE: The relationship one signal has with another. For example, the top half of a cycle is 180 degrees out of phase with the bottom half.

PHOTOCELL: An electronic component that detects light and changes its resistance according to the light's intensity.

PHOTODIODE: A specialized diode that allows light to fall on the semiconductor junction, thus changing its properties.

PHOTOELECTRIC: Another term applied to photocell-related activity.

POTENTIOMETER: A type of resistor that allows its value to be changed. A variable resistor.

POWER: A designation of the available electric potential, as determined by the formula *voltage x current = power.*

POWER SUPPLY: An electronic circuit that produces the necessary power for another circuit or device.

RESISTOR: An electronic component that obstructs (resists) the flow of electricity. They are measured in ohms.

RESOLUTION: This term applies to the quality of a video display. It can involve the number of scan line, the number of pixels, or the distance between the pixels. Also referred to as *dot pitch.*

RETRACE: The process by which an electron gun's scanning beam returns to the top of the screen and retraces the image. Also know as *refreshing.*

SCREEN: Normally refers to the phosphorescent coating on the front of a cathode ray tube. This coating glows when struck by the electron gun's beam. Can also refer to modern LCD/LED displays.

SENSITIVITY: Applies to the selectivity or responsiveness of systems or controls to incoming signals.

SHOCKLEY, W.B.: One of three scientists credited with the 1947 invention of the transistor.

SIGNAL: Any information being passed from one electronic system to another. Transmitted RF is one example of a signal.

SIGNAL INJECTOR: A device that provides a stable signal that can be introduced or injected into a circuit.

SIGNAL TRACER: A device that can track or trace a signal throughout a circuit. The signal can be from a signal injector.

SOLAR CELL: A type of photocell that produces electricity instead of changing resistance; a.k.a., a solar battery.

SOLID-STATE: Refers to electronic components that rely on silicon, germanium, gallium arsenide or other junctions. This is the basis of the transistor, IC, LED, and so forth.

SYNCHRONIZATION (SYNC): The timing signals and pulses that hold a video format together. Without sync, the image appearing on the screen would be gibberish.

SYNC GENERATOR: The electronic circuit that produces precise sync signals/pulses.

TELEVISION (TV): Term for the massive industry of video production and broadcasting coming from the Greek word *tele*, for *far*, and the Latin word *vision*, for *see*. Thus, "television" means "to see far."

TELEVISION CAMERA: Any one of many devices using various imaging components, which convert images into an electronic signal.

TRANSISTOR: A semiconductor component that can be used to amplify signals or as electronic switches. These are the primary element of integrated circuits.

TRANSISTOR TESTER: A device that is used to test and identify transistors and their leads.

TRANSMITTER: The electronic circuit that sends out (for reception) a signal by way of radio waves. The signal usually contains some form of data.

TUBE: Any one of a variety of metal structures, with electrodes connected, enclosed in a glass envelope from which the air has been evacuated. Hence the term *vacuum tube*. In today's consumer equipment, usually the only tubes still in use are CRTs, or picture tubes.

ULTRA-HIGH FREQUENCY (UHF): The radio frequency spectrum from around 400 to 900 megahertz. The greater portion of commercial television (channels 14 to 69) operates within this range, as well as police, fire, and other emergency and commercial radio services.

VERTICAL: Any operation that involves up and down movement as opposed to back and forth.

VERY HIGH FREQUENCY (VHF): The range of frequencies from about 30 to 300 megahertz. This spectrum covers everything from HAM radio to emergency and TV broadcasting.

VIDEO: Pertaining to television or the electronic presentation of images. Also refers to the area of electronics that concerns this process.

VIDEO AMPLIFIER: An electronic circuit that magnifies video signals. Also sometimes referred to as a *differential amplifier*.

WAVEFORM: The shape a signal assumes. Common forms are square, sine and sawtooth or triangular waves.

WAVEFORM MONITOR: A specialized oscilloscope TV technicians use to watch the phase of the various components within a television signal. Components include chrominance information, luminance signals and color burst pulses.

WAVELENGTH: While loosely used to mean *frequency*, this term more appropriately refers to light. Each color has a specific wavelength expressed in nanometers, such as 633 nanometers for red light.

WIRELESS: Applies to sending information without connecting cables. Usually refers to radio-frequency transmissions, although infrared remote controls certainly fit the category.

ZENER: A type of diode that acts as a voltage regulator by restricting the flow of voltage above its rating. For example, a 5-volt zener placed across an input will regulate the voltage of that input to 5 volts.

Index

Index

D

Index

F

G

H

Index

Index

Index

U

V

Index

Your Technology Connection to the Future!

A Bell Atlantic Company

Howard W. Sams

Now You Can Visit Howard W. Sams & Company On-Line: http://www.hwsams.com

Gain Easy Access to:

- **The PROMPT Publications catalog, for information on our *Latest Book Releases*.**
- **The PHOTOFACT Annual Index.**
- **Information on Howard W. Sams' Latest Products.**
- ***AND MORE!***

PROMPT®
PUBLICATIONS

CALL 1-800-428-7267 TODAY FOR THE NAME OF YOUR NEAREST PROMPT PUBLICATIONS DISTRIBUTOR

PROMPT®
PUBLICATIONS

IC Projects: Fun for the Electronics Hobbyist & Technician

Carl Bergquist

IC Projects was written for electronics hobbyists and technicians who are interested in projects for integrated circuits — projects that you will be able to breadboard quickly and test easily. Designed to be fun and user-friendly, all of the projects in this book employ integrated circuits and transistors.

IC projects presents the experienced electronics enthusiast with instructions on how to construct such interesting projects as a LED VU meter, infrared circuit, digital clock, digital stopwatch, digital thermometer, electronic "bug" detector, sound effects generator, laser pointer, and much more!

Electronic Projects
256 pages ♦ Paperback ♦ 6 x 9"
ISBN: 0-7906-1116-3 ♦ Sams: 61116
$21.95 ♦ May 1997

The Video Hacker's Handbook

Carl Bergquist

Geared toward electronic hobbyists and technicians interested in experimenting with the multiple facets of video technology, *The Video Hacker's Handbook* features projects never seen before in book form. Video theory and project information is presented in a practical and easy-to-understand fashion, allowing you to not only learn how video technology came to be so important in today's world, but also how to incorporate this knowledge into projects of your own design. In addition to the hands-on construction projects, the text covers existing video devices useful in this area of technology plus a little history surrounding television and video relay systems.

Video Technology
336 pages + Paperback + 7-3/8 x 9-1/4"
ISBN: 0-7906-1126-0 + Sams: 61126
$24.95 + September 1997

CALL 1-800-428-7267 TODAY FOR THE NAME OF YOUR NEAREST PROMPT PUBLICATIONS DISTRIBUTOR

PROMPT® PUBLICATIONS

Electronic Circuit Guidebook Volume 1, Sensors

Joseph J. Carr

Most sensors are inherently analog in nature, so their outputs are not usable by the digital computer. Even if the sensor is supposedly a digital output design, it is likely that an inherently analog process is paired with an analog-to-digital converter. In *Electronic Circuit Guidebook, Volume 1: Sensors*, you will find information you need about typical sensors, along with a large amount of information about analog sensor circuitry. Amplifier circuits are especially well covered, along with differential amplifiers, analog signal processing circuits and more. This book is intentionally kept practical in outlook. Some topics covered include electronics signals and noise, measurement, sensors and instruments, instrument design rules, sensor interfaces, analog amplifiers, and sensor resolution improvement techniques.

Electronic Theory
340 pages + Paperback + 7-3/8 x 9-1/4"
ISBN: 0-7906-1098-1 + Sams: 61098
$24.95 + April 1997

Electronic Circuit Guidebook Volume 2, IC Timers

Joseph J. Carr

Timer circuits used to be a lot of trouble to build and tame for several reasons. One major reason was the fact that DC power supply variations would cause a frequency shift or slow drift. Part I of this book is organized to demonstrate the theory of how various timers work. This is done by way of an introduction to resistor-capacitor circuits, and in-depth chapters on various TTL and CMOS digital IC devices. Part II presents a variety of different circuits and projects. Some of the circuits include: analog audio frequency meter, one-second timer/flasher, relay and optoisolator drivers, two-phase digital clock and more. *Electronic Circuit Guidebook, Volume 2: IC Timers* will teach you enough that you will not only be able to rework and modify the circuits covered here, but also design a few of your own.

Electronics Technology
240 pages + Paperback + 7-3/8 x 9-1/4"
ISBN: 0-7906-1106-6 + Sams: 61106
$24.95 + August 1997

CALL 1-800-428-7267 TODAY FOR THE NAME OF YOUR NEAREST PROMPT PUBLICATIONS DISTRIBUTOR

Electronic Circuit Guidebook Volume 3, Op Amps

Joseph J. Carr

The operational amplifier is the most commonly used linear IC amplifier in the world. The range of applications for the op amp is truly awesome – it has become a mainstay of audio, communications, TV, broadcasting, instrumentation, control, and measurement circuits. Third in a series covering electronic instrumentation and circuitry, *Electronic Circuit Guidebook, Volume 3: Op Amps* is design to give you some insight into how practical linear IC amplifiers work in actual real-life circuits. Because of their widespread popularity, operational amplifiers figure heavily in this book, though other types of amplifiers are not overlooked. This book allows you to design and configure your own circuits, and is intended to be a practical workshop aid. Some of the topics covered in detail include linear IC amplifiers, ideal operational amplifiers, instrumentation amplifiers, isolation amplifiers, active analog filter circuits, waveform generators, and many more.

Electronics Technology

273 pages + Paperback + 7-3/8 x 9-1/4"

ISBN: 0-7906-1131-7 + Sams: 61131

$24.95 + August 1997

Electronic Circuit Guidebook Volume 4, Electro-Optics

Joseph J. Carr

Electronic Circuit Guidebook, Volume 4: Electro-Optics is mostly about E-O sensors — those electronic transducers that convert light waves into a proportional voltage, current, or resistance. The coverage of the sensors is wide enough to allow you to understand the physics behind the theory of operation of the device, and also the circuits used to make these sensors into useful devices. This book examines the photoelectric effect, photoconductivity, photovoltaics, and PN junction photodiodes and phototransistors. Also examined is the operation of lenses, mirrors, prisms, and other optical elements keyed to light physics.

Electronic Circuit Guidebook, Volume 4: Electro-Optics is intended to teach the physics and operation of E-O devices, then proceed to circuits and methods for actual application of the devices in real situations.

Electronics Technology

416 pages + Paperback + 7-3/8 x 9-1/4"

ISBN: 0-7906-1132-5 + Sams: 61132

$29.95 + October 1997

CALL 1-800-428-7267 TODAY FOR THE NAME OF YOUR NEAREST PROMPT PUBLICATIONS DISTRIBUTOR

PROMPT® PUBLICATIONS

Semiconductor Cross Reference Book Fourth Edition

Howard W. Sams & Company

This newly revised and updated reference book is the most comprehensive guide to replacement data available for engineers, technicians, and those who work with semiconductors. With more than 490,000 part numbers, type numbers, and other identifying numbers listed, technicians will have no problem locating the replacement or substitution information needed. There is not another book on the market that can rival the breadth and reliability of information available in the fourth edition of the *Semiconductor Cross Reference Book.*

Professional Reference
688 pages - Paperback - 8-1/2 x 11"
ISBN: 0-7906-1080-9 - Sams: 61080
$24.95 ($33.95 Canada) - August 1996

IC Cross Reference Book Second Edition

Howard W. Sams & Company

The engineering staff of Howard W. Sams & Company assembled the *IC Cross Reference Book* to help readers find replacements or substitutions for more than 35,000 ICs and modules. It is an easy-to-use cross reference guide and includes part numbers for the United States, Europe, and the Far East. This reference book was compiled from manufacturers' data and from the analysis of consumer electronics devices for PHOTOFACT® service data, which has been relied upon since 1946 by service technicians worldwide.

Professional Reference
192 pages - Paperback - 8-1/2 x 11"
ISBN: 0-7906-1096-5 - Sams: 61096
$19.95 ($26.99 Canada) - November 1996

CALL 1-800-428-7267 TODAY FOR THE NAME OF YOUR NEAREST PROMPT PUBLICATIONS DISTRIBUTOR

PROMPT® PUBLICATIONS

PC Hardware Projects Volume 1

James "J.J." Barbarello

Now you can create your own PC-based digital design workstation! Using commonly available components and standard construction techniques, you can build some key tools to troubleshoot digital circuits and test your printer, fax, modem, and other multiconductor cables.

This book will guide you through the construction of a channel logic analyzer, and a multipath continuity tester. You will also be able to combine the projects with an appropriate power supply and a prototyping solderless breadboard system into a single digital workstation interface!

PC Hardware Projects, Volume 1, guides you through every step of the construction process and shows you how to check your progress.

PROJECT SOFTWARE DISK INCLUDED!

Computer Technology
256 pages - Paperback - 7-3/8 x 9-1/4"
ISBN: 0-7906-1104-X - Sams: 61104
$24.95 - Feb. 1997

PC Hardware Projects Volume 2

James "J.J." Barbarello

PC Hardware Projects, Volume 2, discusses stepper motors, how they differ from conventional and servo motors, and how to control them. It investigates different methods to control stepper motors, and provides you with circuitry for a dedicated IC controller and a discrete component hardware controller.

Then, this book guides you through every step of constructing an automated, PC-controlled drilling machine. You'll then walk through an actual design layout, creating a PC design and board. Finally, you'll see how the drill data is determined from the layout and drill the PCB. With the help of the information and the data file disk included, you'll have transformed your PC into your very won PCB fabrication house!

PROJECT SOFTWARE DISK INCLUDED!

Computer Technology
256 pages - Paperback - 7-3/8 x 9-1/4"
ISBN: 0-7906-1109-0 - Sams: 61109
$24.95 - May 1997

CALL 1-800-428-7267 TODAY FOR THE NAME OF YOUR NEAREST PROMPT PUBLICATIONS DISTRIBUTOR

PROMPT® PUBLICATIONS

Desktop Digital Video

Ron Grebler

Desktop Digital Video is for those people who have a good understanding of personal computers and want to learn how video (and digital video) fits into the bigger picture. This book will introduce you to the essentials of video engineering, and to the intricacies and intimacies of digital technology. It examines the hardware involved, then explores the variety of different software applications and how to utilize them practically. Best of all, *Desktop Digital Video* will guide you through the development of your own customized digital video system. Topics covered include the video signal, digital video theory, digital video editing programs, hardware, digital video software and much more.

Video Technology

225 pages + Paperback + 7-3/8 x 9-1/4"

ISBN: 0-7906-1095-7 • Sams: 61095

$29.95 • June 1997

TV Video Systems

L.W. Pena & Brent A. Pena

Knowing which video programming source to choose, and knowing what to do with it once you have it, can seem overwhelming. Covering standard hard-wired cable, large-dish satellite systems, and DSS, *TV Video Systems* explains the different systems, how they are installed, their advantages and disadvantages, and how to troubleshoot problems. This book presents easy-to-understand information and illustrations covering installation instructions, home options, apartment options, detecting and repairing problems, and more. The in-depth chapters guide you through your TV video project to a successful conclusion.

Video Technology

124 pages - Paperback - 6 x 9"

ISBN: 0-7906-1082-5 - Sams: 61082

$14.95 ($20.95 Canada) - June 1996

CALL 1-800-428-7267 TODAY FOR THE NAME OF YOUR NEAREST PROMPT PUBLICATIONS DISTRIBUTOR

PROMPT®
PUBLICATIONS

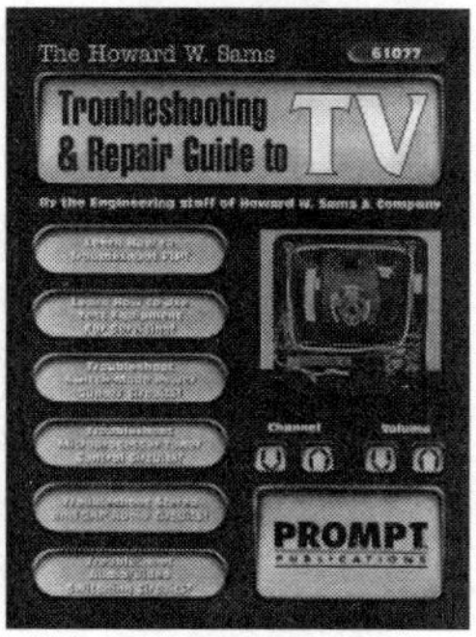

The Howard W. Sams Troubleshooting & Repair Guide to TV

Howard W. Sams & Company

The Howard W. Sams Troubleshooting & Repair Guide to TV is the most complete and up-to-date television repair book available. Included in its more than 300 pages is complete repair information for all makes of TVs, timesaving features that even the pros don't know, comprehensive basic electronics information, and extensive coverage of common TV symptoms.

This repair guide is completely illustrated with useful photos, schematics, graphs, and flowcharts. It covers audio, video, technician safety, test equipment, power supplies, picture-in-picture, and much more. *The Howard W. Sams Troubleshooting & Repair Guide to TV* was written, illustrated, and assembled by the engineers and technicians of Howard W. Sams & Company.

Video Technology
384 pages - Paperback - 8-1/2 x 11"
ISBN: 0-7906-1077-9 - Sams: 61077
$29.95 ($39.95 Canada) - June 1996

The In-Home VCR Mechanical Repair & Cleaning Guide

Curt Reeder

Like any machine that is used in the home or office, a VCR requires minimal service to keep it functioning well and for a long time. However, a technical or electrical engineering degree is not required to begin regular maintenance on a VCR. *The In-Home VCR Mechanical Repair & Cleaning Guide* shows readers the tricks and secrets of VCR maintenance using just a few small hand tools, such as tweezers and a power screwdriver.

This book is also geared toward entrepreneurs who may consider starting a new VCR service business of their own. The vast information contained in this guide gives a firm foundation on which to create a personal niche in this unique service business. This book is compiled from the most frequent VCR malfunctions Curt Reeder has encountered in the six years he has operated his in-home VCR repair and cleaning service.

Video Technology
222 pages - Paperback - 8-3/8 x 10-7/8"
ISBN: 0-7906-1076-0 - Sams: 61076
$19.95 ($26.99 Canada) - April 1996

CALL 1-800-428-7267 TODAY FOR THE NAME OF YOUR NEAREST PROMPT PUBLICATIONS DISTRIBUTOR

PROMPT® PUBLICATIONS

Basic Digital Electronics

Alvis J. Evans

Explains digital system functions and how digital circuits are used to build them! Digital — what does it mean? Why is it that electronic systems are being designed using digital electronic circuits? Find the answer to these questions and more, as you learn the difference between analog and digital systems, the functions required to design digital systems, the circuits used to make decisions, code conversions, data selections, adding and subtracting, interfacing and storage, and the circuits that keep all operations in time and under control.

Learn about logic circuits, flip-flops, registers, multivibrators, counters, 3-state bus drivers, bidirectional line drivers and receivers, and more using easy-to-read, easy-to-understand explanations coupled with detailed illustrations.

Electronic Theory
192 pages + Paperback + 8-1/2 x 11"
ISBN: 0-7906-1118-X + Sams: 61118
$19.95 + April 1997

Electronic Projects for the 21st Century

John Iovine

If you are an electronics hobbyist with an interest in science, or are fascinated by the technologies of the future, you'll find Electronic Projects for the 21st Century a welcome addition to your electronics library. It's filled with nearly two dozen fun and useful electronics projects designed to let you use and experiment with the latest innovations in science and technology — innovations that will carry you and other electronics enthusiasts well into the 21st century!

Electronic Projects for the 21st Century contains the expert, hands-on guidance and detailed instructions you need to perform experiments that involve genetics, lasers, holography, Kirlian photography, and more. Among the projects are a lie detector, an ELF monitor, air pollution monitor, pinhole camera, laser power supply for holography, synthetic fuel, and an expansion cloud chamber.

Electronic Projects
256 pages + Paperback + 7-3/8 x 9-1/4"
ISBN: 0-7906-1103-1 + Sams: 61103
$19.95 + June 1997

CALL 1-800-428-7267 TODAY FOR THE NAME OF YOUR NEAREST PROMPT PUBLICATIONS DISTRIBUTOR

PROMPT® PUBLICATIONS

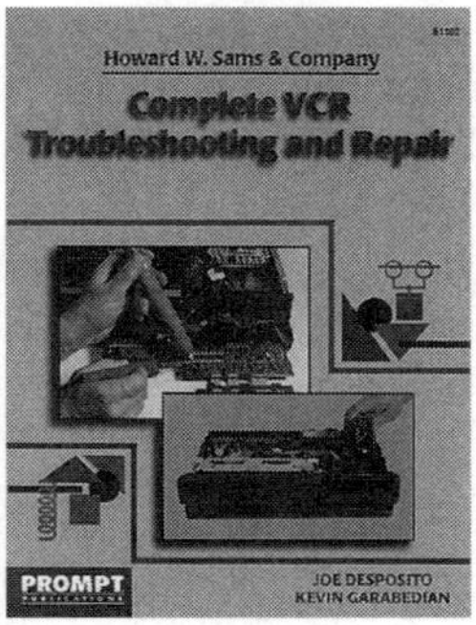

Howard W. Sams Complete VCR Troubleshooting & Repair

Joe Desposito & Kevin Garabedian

Complete VCR Troubleshooting and Repair contains sound VCR troubleshooting procedures beginning with an examination of the external parts of the VCR, then narrowing the view to gears, springs, pulleys, belts, and other mechanical parts. This book also shows how to troubleshoot tuner/demodulator circuits, audio and video circuits, display controls, servo systems, video heads, TV/VCR combination models, and more.

This book also contains nine VCR case studies, each focusing on a particular model of VCR with a specific problem. The case studies guide you through the repair from start to finish, using written instruction, helpful photographs, and Howard W. Sams' own *VCRfacts*® schematics.

Video Technology
184 pages s Paperback s 8-1/2 x 11"
ISBN: 0-7906-1102-3 s Sams: 61102
$29.95 s March 1997

Howard W. Sams Computer Monitor Troubleshooting & Repair

Joe Desposito & Kevin Garabedian

Computer Monitor Troubleshooting & Repair makes it easier for any technician, hobbyist or computer owner to successfully repair dysfunctional monitors. Learn the basics of computer monitors with chapters on tools and test equipment, monitor types, special procedures, how to find a problem and how to repair faults in the CRT. Other chapters show how to troubleshoot circuits such as power supply, high voltage, vertical, sync and video.

This book also contains six case studies which focus on a specific model of computer monitor. Using carefully written instructions and helpful photographs, the case studies guide you through the repair of a particular problem from start to finish. The problems addressed include a completely dead monitor, dysfunctional horizontal width control, bad resistors, dim display and more.

Troubleshooting & Repair
308 pages + Paperback + 8-1/2 x 11"
ISBN: 0-7906-1100-7 + Sams: 61100
$29.95 + July 1997

CALL 1-800-428-7267 TODAY FOR THE NAME OF YOUR NEAREST PROMPT PUBLICATIONS DISTRIBUTOR

PROMPT® PUBLICATIONS

Home Security Projects

Robert Gaffigan

Home Security Projects presents the reader with many projects about home security, safety and nuisance elimination that can easily be built in the reader's own home for less than it would cost to buy these items ready-made. Readers will be able to construct devices that will allow them to protect family members and electrical appliances from mishaps and accidents in the home, and protect their homes and belongings from theft and vandalism.

This book shows the reader how to construct the many useful projects, including a portable CO detector, trailer hitch alignment device, antenna saver, pool alarm, dog bark inhibitor, and an early warning alarm system. These projects are relatively easy to make and the intent of *Home Security Projects* is to provide enough information to allow you to customize them.

Projects
256 pages + Paperback + 6 x 9"
ISBN: 0-7906-1113-9 + Sams: 61113
$19.95 + September 1997

Security Systems for Your Home & Automobile

by Gordon McComb

Security Systems is about making homes safer places to live and protecting cars from vandals and thieves. It is not only a buyer's guide to help readers select the right kind of alarm system for their home and auto, it also shows them how to install the various components. Learning to design, install, and use alarm systems saves a great deal of money, but it also allows people to learn the ins and outs of the system so that it can be used more effectively. This book is divided into eight chapters, including home security basics, warning devices, sensors, control units, remote paging automotive systems, and case histories.

Gordon McComb has written over 35 books and 1,000 magazine articles which have appeared in such publications as *Omni* and *PC World.* In addition, he is the coauthor of PROMPT® Publication's *Speakers for Your Home and Auto.*

Projects
130 pages + Paperback + 6 x 9"
ISBN: 0-7906-1054-X + Sams: 61054
$16.95 + July 1994

CALL 1-800-428-7267 TODAY FOR THE NAME OF YOUR NEAREST PROMPT PUBLICATIONS DISTRIBUTOR